post-structuralist
geography

post-structuralist geography

a guide to relational space

jonathan murdoch

First published 2006

SAGE Publications Ltd
1 Oliver's Yard
55 City Road
London EC1Y 1SP

SAGE Publications Inc.
2455 Teller Road
Thousand Oaks, California 91320

SAGE Publications India Pvt Ltd
B-42, Panchsheel Enclave
Post Box 4109
New Delhi 110 017

British Library Cataloguing in Publication data

A catalogue record for this book is available from the British Library

ISBN 0 7619 7423 7
ISBN 0 7619 7424 5 (pbk)
ISBN 978-0-7619-7424-6
Library of Congress Control Number: 2005925439

Typeset by C&M Digitals (P) Ltd., Chennai, India
Printed on paper from sustainable resources
Printed in Great Britain by Athenaeum Press, Gateshead

To Mara, Carlotta and the beautiful golden boy

Contents

Acknowledgements

From a post-structuralist perspective, books, even those that are single-authored, constitute 'collectives' – that is, they draw together writings, readings, references, quotations, data, thoughts, experiences, discussions, arguments, and many other things besides. From the text, then, it is possible to trace outwards multiple 'lines of flight' (as the post-structuralist philosopher Gilles Deleuze might put it) from text to context. Clearly, some of these 'lines of flight' are easier to follow than others. Quotations are referenced and references show how connections have been established with other texts so that a discursive network eventually comes into view. This network is rendered reasonably transparent by academic writing conventions. However, other relationships are hidden within the various arguments or descriptions that are mobilized in the text; they are embedded into the narrative in ways that make them difficult to discern. It is these 'hidden' or personal relations that need to be 'disembedded' and brought out into the open.

In acknowledging the personal relations that have decisively shaped the text that follows I'd like to begin at the beginning. I was very fortunate to have a number of stimulating teachers during my undergraduate and postgraduate studies in Aberystwyth in the early- to mid-1980s, in particular Graham Day (who taught me how to think sociologically) and Karel Williams (who first introduced me to Foucault's take on things). From Aberystwyth I moved to London to work on a UK Economic and Social Research Council [ESRC] project run by Philip Lowe, Terry Marsden and Richard Munton. Under their supervision I was not only introduced to the broader geographical research community but was given ample opportunity to wander off in many theoretical directions. I thus developed a new interest in the relationship between sociology and geography and this led me into an encounter with actor-network theory. Philip, Terry and Richard were all supportive of my early attempts to make this theory geographically useful.

In the early 1990s, I moved to the University of Newcastle to take up a Fellowship at the Centre for Rural Economy. Again, Philip Lowe helped to create an open and engaging working environment where many innovative ideas about nature, food and rurality were discussed and elaborated and, again, I acknowledge my debt to him. Others based in Newcastle at the same time also contributed to a vibrant intellectual community, especially Ash Amin, Alistair Bonnet, Nina Laurie, Simon Marvin and Neil Ward. I then moved to my current institutional home in Cardiff at the School of City and Regional Planning. Over the years this has proved an extremely supportive context for my work even though much of what I do holds only a tangential relationship to planning. Many colleagues have provided valuable encouragement but in particular I would like to thank Jeremy Alden (for a warm welcome to Cardiff in the first instance), Kevin Morgan (for boundless enthusiasm on a variety of research and other topics) and Terry Marsden (for constant and steadfast support).

Much of the material presented in this book has been gestating for some considerable time. However, two of the chapters in the second half of the book derive from recent research projects, notably, a piece of ESRC-funded research on 'Environmental Action and the Policy Process' conducted in England between 1999 and 2001 with Philip Lowe and Andrew Norton, and some unfunded work on Slow Food conducted in Italy between 1999 and 2001 with Mara Miele. These colleagues have provided considerable assistance in analysing the research material, although I take full responsibility for the views expressed on these topics in what follows. I am also indebted to Simone Abram for helping me to develop my thinking on the nature of planning over the course of two ESRC projects which examined how technical and political forms of decision making interact within planning processes. Likewise, my colleagues Richard Cowell and Neil Harris have set me straight on a number of planning issues over the years, especially those considered in Chapter 6.

As well as this support from institutional colleagues I have benefited over the years from occasional conversations with theoretical 'fellow travellers'. Mention should be made of Noel Castree, Judy Clark, Paul Cloke, Julia Cream, David Goodman, Andy Pratt and Sarah Whatmore. These and other theoretically-minded colleagues have provided a great deal of encouragement for the post-structuralist explorations that follow. Finally, a number of people have provided valuable assistance in getting the final manuscript into shape. I would like to thank Tom Garne (for help with the illustrations), Joek Roex (for extensive help with editing and for his judgements on when infinitives could and could not be split), Robert Rojek (for guidance and patience, two essential qualities in any editor) and Vanessa Harwood (for steering the book through production in a relatively painless fashion).

I, day.

I, crinoid. I, bristle-worm.

I, Goliath beetle. I, moss. I, orb spider.

I, magnolia. I, wobbegong. I, spurge hawk moth.

I, thallus. I, water. I, cyathozooid. I, horseshoe crab.

I, pink flower mantis. I, swallowtail. I, prairie dog. I, locust.

I, capercailzie. I, manta ray. I, rattlesnake. I, sycamore. I, manatee.

I, honeybee. I, saddleback. I, lamprey. I, coelocanth. I, okapi. I, salmon.

I, snapper. I, fire. I, anolis lizard. I, orchid. I, tree frog. I, iguana. I, ant.

I, skink. I, sidewinder. I, archaeopteryx. I, hoatzin. I, heron. I, king penguin.

I, saddle-billed stork. I, bird of paradise. I, cormorant. I, bower bird. I, macaque.

I, cassowary. I, darter. I, armadillo. I, marmoset. I, bluebell. I, humpbacked whale.

I, possum. I, spruce. I, sequoia. I, gibbon. I, zebra. I, wild dog. I, chimpanzee. I, sloth.

I, wallaby. I, Washingtonia. I, pearly nautilus. I, heron. I, tupaia. I, ptarmigan. I, skunk.

I, brachiopod. I, dolphin. I, bear. I, dachshund. I, koala. I, chiwawa. I, vampire bat. I, vole.

I, silverfish. I, hornet. I, camelia. I, panther. I, chipmunk. I, primrose. I, tyrannosaurus rex.

I, crown-of-thorns starfish. I, argus pheasant. I, human. I, chameleon. I, stick insect. I, bract.

I, thorny devil. I, Portuguese man o'war. I, moon rat. I, flying fish. I, blue-footed booby.

I, ferret. I, sparrow. I, lion. I, kmodo dragon. I, thistle. I, yucca moth. I, stratocumulus.

I, cicada. I, apatosaurus. I, kakapo. I, dipper. I, tapir. I, gurnard. I, segmented worm.

I, Venus' flower basket. I, tiger. I, stork. I, microbe. I, mallee fowl. I, hummingbird.

I, macaw. I, dragonfish. I, hawkmoth. I, jaguar. I, impala. I, caecilian. I, sea-horse.

I, swimbladder. I, orang-utan. I, Douglas fir. I, pipa toad. I, woodpecker. I, air.

I, wryneck. I, baboon. I parakeet. I, horse. I, hedgehog. I, protistan. I, weasel.

I, great crested grebe. I, mushroom. I, hydrogen. I, limpet. I, flying fox.

I, hammerhead. I, mudskipper. I, brittle star. I, pig. I, gecko. I, gannet.

I, axolotl. I, tortoise. I, solenodon. I, python. I, parrot. I, toucanet.

I, monkey-puzzle. I, oxygen. I, hyena. I, gourami. I, springbok.

I, moa. I, ichthyosaur. I, daffodil. I buzzard. I, impala.

I, numbat. I, albatross. I, cat. I, liverfluke. I, slug.

I, lancelet. I, hesperornis. I crocodile.

I, velvet monkey. I, salamander.

I, night.

Robert Crawford, 'Bio'. (This poem first appeared in *The London Review of Books*, www.lrb.co.uk)

1

Post-structuralism and relational space

Geography is the study of relations between society and the natural environment. Geography looks at how society shapes, alters and increasingly transforms the natural environment, creating humanised forms from stretches of pristine nature, and then sedimenting layers of socialisation, one within the other, one on top of the other, until a complex natural-social landscape results. Geography also looks at how nature conditions society, in some original sense of creating the people and raw materials which social forces 'work up' into culture, and in an ongoing sense of placing limits and offering material potentials for social processes [...] The 'relation' between society and nature is thus an entire system, a complex of interrelations [...] Thus, the synthetic core of geography is a study of nature–society relations. (Peet, 1998)

Introduction

Human geography, like all the social science disciplines, has been profoundly affected by post-structuralist theory. Post-structuralist influences can be discerned in almost all aspects of geographical endeavour, in particular the study of cultures, economies and natures – three key areas for human-geographical inquiry. Moreover, post-structuralism has affected not just *what* geographers study but *how* they study. In the wake of post-structuralism's incursion, geographers arguably investigate a broader range of socio-spatial phenomena than was the case previously and do so using innovative research methods (drawn mainly from the qualitative wing of research practice). Writing styles have also changed, with less attention now paid to the communication of scientific rigour and with rather more emphasis placed on the aesthetic and inventive character of geographical discourses and texts (Barnes and Duncan, 1991).[1]

In a whole host of ways, then, post-structuralism has made itself felt within the geographical domain. And yet the full extent of post-structuralism's influence is not easy to gauge. In part, uncertainty about the status of post-structuralism arises from the apparently controversial nature of the changes prompted by any engagement with post-structuralist theory. More playful writing styles, critics argue, do little to bolster public faith in human geography as an accurate and reliable means of gaining knowledge about society (and nature). Likewise, the new research foci and associated research methods are thought to undermine the notion that human geography is a rigorous and meticulous enterprise; rather, the discipline now seems subject to constant and whimsical changes of intellectual fashion. Uncertainty also arises because post-structuralism itself is hard to define; thus, there appear to be many post-structuralisms, each accompanied by its own particular set of theoretical and empirical concerns. Post-structuralism's endless variety means it becomes hard to draw any clear line or boundary around post-structuralist theory or its influence. Thus, while there may be a consensus amongst geographers that post-structuralism has changed the nature of geography, there is likely to be very little agreement on any benefits accruing from that change.

This book has not been written with the intention of resolving all disagreements over the status of 'post-structuralist geography'; rather, it comprises a selective and partial reading of the post-structuralist literature. For instance, very little of what follows is directly concerned with those important strands of post-structuralist work that revolve around 'identity' (this includes post-structuralist feminism as well as work on ethnicity, post-colonialism and sexuality). The analysis instead concentrates on a small number of key post-structuralist thinkers and examines how these thinkers might help geographers to re-conceptualize the spatial realm in ecological and relational terms. In other words, the primary concern of the book is the nature of space, in particular, the relationship between *spatial* processes and *social* processes. It is argued that this analytical focus allows us to appreciate both the contribution post-structuralism can make to geography and the contribution geography can make to post-structuralism.

Particular attention is therefore paid to theorists whose main concern is the *heterogeneous make-up* of spatial formations, that is, the focus is upon forms of post-structuralism which prioritize the materiality of space and the way humans are embedded within spatialized materialities. This conceptual spotlight means that the more 'social' and 'textual' aspects of post-structuralist theory are largely ignored. In defence of the analytical strategy adopted here, it is argued that these 'materialist' post-structuralisms arguably hold great significance for the study of those nature–society interactions that Richard Peet (in the quotation provided at the head of this chapter) sees as the 'synthetic core' of geography. Thus, in what follows, versions of post-structuralism that take a broadly 'ecological' approach constitute the main focus of enquiry. One main aim of the book, then, is to show that post-structuralist theory and human geography have much in common: they both examine nature–society interactions and concern themselves with the (spatial) consequences of these interactions.

Another feature common to post-structuralism and human geography is an interest in 'relationalism'. Any interaction between a people and a 'thing' must also be seen as a relation between the people and the 'thing'. As post-structuralists and geographers begin to look closely at spatially-situated interactions so they begin to recognize that there are many differing kinds of relations running through and around given spatial locations. As the quotation from Peet again makes clear, geographers have long known that interactions between society and nature are relational in character. However, the sub-disciplinary division between *human* and *physical* geography has often prevented the discipline from entering fully into a relational mode of thinking. Thus, human geographers have tended to focus upon *social* relations while physical geographers have attended to *natural* relations. The gap between the two sub-disciplines has never been fully closed, despite recent efforts to forge a closer alignment (cf. Massey, 1999a; Lane, 2001).

Against this background, it will be argued that post-structuralist theory brings significant opportunities for the further development of relational approaches. In particular, it will be shown that post-structuralism's interest in *heterogeneous* relations – that is, in mixtures of the natural and social and the human and the non-human – can help human geographers to reach across the human–physical divide (see Murdoch, 1997; Massey, 1999a; Whatmore, 2002). Moreover, it is argued that a concern for heterogeneity easily translates into a concern for space as these sets of relations necessarily bring together social and natural entities within specific spatial formations (Thrift, 1996). A concern for heterogeneous relations can, then, be given impetus by the infusion of post-structuralist thinking into geographical theory. The impacts of this is 'infusion' will constitute the main focus of the chapters that follow.

However, before we embark on the study of geographical spatialities and post-structuralist relations, we first need to define a little more clearly just what is meant by 'post-structuralism'. Thus, in this chapter, we consider post-structuralism's relationship to 'structuralism'. Having traced post-structuralism's emergence onto the theoretical scene we then go on to assess the impact of post-structuralist thinking upon human geography and briefly review the ways in which post-structuralist analysis has affected perceptions of the spatial realm. Two main impacts are identified: first, a new attention to differences in spatial identifications; second, a new interest in processes of spatial emergence. Both of these impacts are briefly assessed.

While the main purpose of this introductory chapter is to clarify the meaning of the term 'post-structuralist geography', it also provides a way of moving quickly to those strands of post-structuralist theory that constitute the main focus of the book (notably in Chapters 2, 3 and 4). Thus, some time is spent looking at the impact of more 'ecological' post-structuralisms on human geography. It is shown that work conducted under the heading of 'post-structuralism' has challenged some basic geographical assumptions about the composition of the spatial realm by highlighting the embodied and biological nature of social being in space. The ecological approach shades neatly into the relational approach, that is, it emphasizes how heterogeneous relations link social actors into particular spatial domains. In the latter part of the chapter, post-structuralist encouragement for relational thinking is considered. It is argued that relationalism opens geography up to dynamic and complex processes of change. Thus, it is shown that space can no longer be seen as simply a 'container' of heterogeneous processes; rather, space is now thought to be something that is (only provisionally) stabilized out of such turbulent processes, that is, it is made *by* heterogeneous relations. This insight sets the theoretical scene for the chapters that follow, notably those presented in Part 1 of the book.

From structuralism to post-structuralism

The prefix 'post' often leads to the assumption that post-structuralism has much in common with 'post-modernism' – that is, it is yet another attempt to delineate specific features of contemporary society on the basis of some assumed historical shift from one distinct social condition to another. But post-structuralism does not quite work in this way, for it has a rather precise meaning: it refers to philosophical and social theories that come in some sense 'after' structuralism. This indicates that post-structuralism is a term that only really gains any meaning by reference to *theory*, especially as it is employed in philosophy and the social sciences. Yet, while this observation may help in demarcating post-structuralism from much broader but parallel terms such as 'post-modernism', we need to recognize that, for the general reader, this clarification is really no clarification at all. This is because the term 'structuralism' is likely to be shrouded in as much mystery as 'post-structuralism'. Thus, in order to gain some basic understanding of post-structuralism, it is first necessary to say something about structuralism. We can then go on to show how and why post-structuralism emerged in the ways that it did and say something about the nature of its concerns.

Structuralism as an intellectual movement emerged during the early years of the twentieth century with the work of the French linguist Ferdinand de Saussure. In his posthumously published *Course in General Linguistics* (1986 [1916]), Saussure laid the foundations for a structuralist approach to the study of language. In Saussure's view, the relationship between given words and given objects is purely arbitrary – that is, there is no necessary relation between the word 'dog' and that object which comes to be demarcated as 'dog'. Language cannot be seen as simply a reflection of the world 'out there' (beyond language); rather it should be understood as a system of signs which functions to *signify*. In order to understand language, it therefore follows that we need to understand the structure of the symbolic system and the relationships between the various elements that make it up. In Saussure's account, the system can be understood by 'mapping' it out in synchronic (as opposed to diachronic) fashion. Once the system has been mapped, it can be subject to 'depth analysis': this would aim to reveal the underlying structures of any language system (which Saussure called 'langue', in distinction to everyday language uses which he termed 'parole'). As Philip Smith explains,

Saussure's accomplishment is universally acknowledged as formidable. By emphasizing the arbitrary nature of language and its internal structure and logic, he showed that it was a *sui generis* phenomenon which could not be explained away as a mere reflection of reality itself or as an ideology. This is because meaning is generated *within* the linguistic system via a system of differences. Aside from demonstrating the autonomy of language, this point can be applied more widely to underpin the autonomy of any conventional signifying system. (2001: 99)

Saussure's structuralist approach became enormously influential within the discipline of linguistics. Yet, its impact was arguably greatest in the social sciences. In particular, it was taken up by the leading anthropologist, Claude Lévi-Strauss, in the middle years of the twentieth century. In Lévi-Strauss's hands, structuralism came into its own and his anthropological form of structuralism now stands as perhaps the clearest example of the genre (Edith Kurzweil (1980), for instance, calls Lévi-Strauss 'the father of structuralism').

Lévi-Strauss sought to use structuralist analysis as a means of comprehending the underlying structures of diverse human cultures. He first applied the approach in his (1969 [1949]) book *The Elementary Structures of Kinship*. In this work he brought together a vast array of ethnographic material in order to show that beneath the layers of cultural diversity lies some kind of underlying and determining structure:

Essentially, kinship systems could be reduced to a limited pool of types. These in turn were underpinned by an equally limited set of 'superficially complicated and arbitrary rules [that] may be reduced to a smaller number'. The most important of these concerned issues of descent (e.g. who could belong to a given clan) and marriage (e.g. who was allowed to marry whom). Even more fundamental than kinship rules was the incest taboo, a universal prohibition which Lévi-Strauss saw as being the point of origin for cultural life. Thanks to the incest taboo, he claimed, people were forced to become sociable in the search for mates. (Smith, 2001: 103)

By identifying the generative mechanisms of kinship systems – descent, marriage and incest – Lévi-Strauss believed he could explain the complex details of social and cultural life. Thus, in his view, the aim of structuralist anthropology was to go beneath the surface to focus on 'mechanisms' rather than people or behaviour.

The approach was taken further in later work, notably his analysis of myths. Again, Lévi-Strauss attended to the elements that work to generate societal myths. He sought to break the myths down into their constituent units, and he endeavoured to show how combinations of these units could be understood in terms of binary oppositions between, say, life/death, nature/culture and raw/cooked. He argued that these oppositions help to explain localized variation in myth-making so that their use can uncover the underlying order in 'universal' myths.

In undertaking this analysis, Lévi-Strauss effectively read *across* texts (myths) in order to establish a common set of interlocking elements. Tudor summarizes the formalized nature of the approach:

A Lévi-Straussian analysis [...] demanded that the analyst identified the units from which texts were constructed, did so largely in isolation from the actual reading practices of consumers of those texts, and arrived at an account of 'meaning' by examining the formal combinations and permutations of these units across the (trans-cultural) corpus of texts. Such an approach is 'formalist' in several senses. First and most obviously, it focuses on the formal patterning of cultural materials across the whole set of artefacts, treating this as revealing the most 'significant meanings' which texts carry. In doing that, however, it abstracts texts from their culture, reifying revealed form. The texts come to carry meaning in consequence of the structures that the analysis uncovers, a process which functions quite independently of the social agents who make and use culture. In other words, both the social and the individual recedes into the background of such an analysis – the 'forms' themselves provide sufficient grounds for credible interpretative conclusions. (1999: 69)

Lévi-Strauss synthesized an astonishingly large number of cultures and texts in his analyses. Yet, his attempts to discern the 'generative mechanisms' underlying these cultures and texts met with only limited success and later generations of anthropological researchers eschewed his structuralism in favour of more ethnographic approaches (Clifford and Marcus, 1986).

BOX 1.1

Philip Smith (2001: 97–8) describes the core features of structuralist analysis thus:

- Depth explains surface. For the structuralists the seeming chaotic and unpredictable character of social life is something of an illusion: 'Beneath the level of perplexing and unique events are hidden generative mechanisms'.
- This depth is structured. These 'generative mechanisms' are ordered, organized and patterned and are made up of a limited number of elements. These limited elements combine to 'generate' surface phenomena such as events, actions, beliefs, cultures and so forth.
- The analyst is objective. 'Structuralists see themselves as detached scientific observers who are discovering some kind of truth that is not apparent to social actors'.
- Culture is like language. Culture is systemic and should be analysed as a total social form in which various elements combine to generate meaning and to stimulate action. This 'systemic' view means that cultures and their associated societies are inevitably closed, separated off from alternative cultures and societies.
- Structuralism is 'beyond' humanism. The human subject is of little consequence in this approach for meanings and actions arise not from individuals but from the 'generative mechanisms' which underlie social formations. Thus, any investigation of society should focus on the mechanisms rather than social actors. 'The major focus is on the role and workings of the cultural system, rather than on the consciousness and genius of the individual human agent'.

Despite widespread criticisms of the structuralist method, Lévi-Strauss's work on myth was subsequently picked up by literary scholars in the 1950s and 1960s. In particular, Roland Barthes published a number of influential articles on modern 'mythologies'. In these articles (which are collected in Barthes, 1993) he sought to explain the 'real' meaning of such everyday items as domestic cleaning products or skin-care creams. In studying the advertising that routinely surrounds such artefacts, Barthes discerned a whole range of political, cultural and social associations that combine to give them 'meaning'. In an essay entitled 'Myth Today' published at the end of *Mythologies*, Barthes draws out some general lessons from his analysis and in so doing displays his debt to both Saussure and Lévi-Strauss, for instance, through his continued use of mythological codes. Yet, in this essay, Barthes also makes clear the need to combine the scientific study of texts with more sociological understandings of meaning and practice. For Barthes, a full analysis of modern mythologies can only be undertaken once deeper structures are aligned with specific social, cultural and economic forms (for example, capitalism and capitalist ideology).

In *Mythologies* Barthes begins to move structuralism away from the over-formalized approach developed by Lévi-Strauss: instead, he combines abstract analysis of underlying structures with more impressionistic assessments of social context. This move is taken further in his later work, notably *S/Z*, which comprises a detailed analysis of the story 'Sarrasine' by Balzac (Barthes, 1975). In the study, Barthes identifies five codes which he claims facilitate analysis of Sarrasine's narrative structure. Yet, at the same time as he develops a structuralist reading, Barthes makes clear that there is no *single* and *definitive* meaning embedded in the text. As Smith (2001: 112) points out, Barthes suggests that 'the complex codes in operation overlap with each other in wild and unpredictable ways. There is an excess of meaning. The codes open up possibilities for alternative interpretations, and so there is a role for the reader in making sense of the text'.

The appearance of the reader as an active agent in the generation of meaning is an important step forward, for it infers that meaning cannot be apprehended simply through the scientific study of formal structures. As Connelly (1999: 58) puts it, 'some elements in the existing [textual] code [...] must be modified if space is to be opened for something new to emerge'. Thus, meaning should be seen as arising from a relationship between the reader *and* the text. Moreover, this relationship is non-deterministic – that is, the reader is not simply configured by the codes in the story but is an active agent in the process of meaning generation. In other words, relations between subjects and objects are central to the functioning of social or cultural systems.

The shift in *S/Z* from the scientific analysis of formal structures to an engagement with multiple meanings is often seen as the first major move from structuralism to post-structuralism (Belsey, 2002). Following Barthes, it was possible to argue that cultures, societies and texts are open to diverse interpretations for there is no longer one single deterministic explanation being generated by underlying mechanisms. Now meanings can proliferate in perhaps contradictory ways: no longer are texts, cultures or societies to be seen as closed systems in which elements are structurally locked together in a timeless sequence; rather, systems are open, dynamic and fluid. In order to study such non-linear, unstructured systems something less than formal analysis is required: theorists need to be alert to change, divergence and difference and these qualities can only be fully appreciated if analysts remain somewhere near the *surface* of the phenomena under investigation. In other words, 'depth analysis' gives way to 'breadth analysis'.

For the post-structuralists who follow in the footsteps of Barthes the first task is to engage with the multiplicities of meaning that inevitably emerge from the relationship between texts and their readers. In the work of Derrida, this relationship gives rise to proliferations of meaning as supplements and traces branch off from the main arguments of the text (Derrida, 1998 [1967]). As Judith Butler (2004: 32) explains, Derrida's notion of 'reading' suggests that 'our ability to understand relies on our capacity to interpret signs. It also presupposes that signs come to signify in ways that no particular author or speaker can constrain in advance through intention'. In this context, 'ambiguity, uncertainty and instability always seem to haunt efforts to generate the certain and the definitive' (Smith, 2001: 131). Now, any efforts to establish simple or foundational truths (in the style of structuralism) are doomed to failure as 'entirely clear and hence infallible intellectual insights are found to contain questionable assumptions [and] allegedly pure sense data turn out to embody culturally relative interpretative frameworks' (Gutting, 2001: 295). Again, there is an excess, an overflow, an escape of meaning. The existence of this 'excess' means that efforts to close down interpretation, to force a single narrative onto multiple perspectives, are now rendered problematic, even illegitimate. Any such efforts should be seen not as simple reflections of underlying ('true') generative mechanisms but as historically situated interpretations that come laden with their own perspectival limitations.

Post-structuralism's concern for the multiplicity of meaning seamlessly runs into a concern for multiple identities: as Spivak (1992: 187) puts it, with the emergence of post-structuralism there arises a need to attend to language 'as the production of agency'. Agency emerges from an interaction between symbolic systems and localized practices of meaning generation. Yet, despite the scope for agency, multiplicity is not endless, in part because efforts are constantly being made to close some readings down so that others can gain prominence. One illustration of this process of 'closing down' is provided by Derrida in his examination of readings that privilege male perspectives over female perspectives. He refers to this as 'phallocentrism', a narrative that routinely presents masculine traits as obviously superior to female traits. In a similar vein, post-structuralist feminists such as Luce Irigaray assert the need to establish female traits on new terms so that any subordination to masculinity can be avoided (Irigaray, 1985). These traits can be re-constructed from 'residues' that elude masculine domination in the same way that multiple readings of the text can be based on the traces that escape the 'master' narrative. A similar approach can be taken to non-White or non-Western identifications (Bhabba, 1994). The cracks in the narrative permit alternative meanings, desires, wants and needs to be identified and asserted. All these may run counter to, or may work to undermine, the dominant narrative structure. This leads Lyotard (1988), amongst others, to suggest that *difference* is now more significant than *unity* for the ethical 'we' that had previously worked to draw all identities into the realm of 'sameness' has been undermined by the splintering of meaning into various situated practices of meaning generation. In the wake of the shift from structuralism to post-structuralism, any efforts to corral the multitude must be seen as either exclusionary or coercive or both. Following the deconstruction of the single narrative, diversity and multiplicity must, indeed, should reign.

In short, post-structuralism comes *after* structuralism. Once we recognize the temporal development of theory within philosophy and the social sciences, we can gain a more precise understanding of what 'post-structuralism' as a concept and as a body of theory might mean. It refers, in the main, to the multiple meanings and modes of identification that emerge from the constitution of relations within texts and within cultures. In this respect, it differs sharply from structuralism's concern to discover underlying truths about texts and cultures. Yet, post-structuralism should not be seen simply as a clean break from structuralism. Many post-structuralist authors retain an obvious debt to structuralism (Roland Barthes provides one example, Foucault – as we shall see in Chapter 2 below – provides another), and they continue to work with many of the core precepts of the earlier theory, notably the idea that meaning is generated not by knowing individuals but by sets of textual, cultural and social relations – as Belsey (2002: 72) says of post-structuralism: 'the subject is an effect of culture, a result of the circulation of meanings in the symbolic order, rather than their origin'.

BOX 1.2

Some core features of post-structuralist analysis:

- Meaning and action must be set in a context of extensive relations. Post-structuralism retains the structuralist concern for 'systems' rather than individuals – it thus remains anti-humanist. However, it does not believe these systems can be understood through the discernment of underlying structure. Rather, it focuses on the extensive nature of systemic relations.
- Systems are now 'open' rather than 'closed'. Thus, meanings and actions cannot be seen as simply manifestations of underlying structures – they proliferate in complex and unexpected ways, depending on the relations established between subjects and objects within the system.

Continued

- Relations between subjects and objects are subject to contestation. Just as there are struggles to establish the meanings of texts, there are struggles to establish identities of various kinds. These struggles and contestations easily become political, subject to plays of power.
- There is an interplay between systemic relations and struggles over meaning and identity. As Barthes showed in his work on modern mythologies, in capitalist society some meanings become easily privileged over others. However, these meanings are not fixed for all time in the broader system (as structuralists tend to believe): they are likely to be changed as new interpretations emerge and as new identifications come into being.
- Subjectivity is 'decentred'. Meaning and identity develop as part of relational systems. Thus, identity is decentred across such systems. This means that subjectivity is fragmented as subjects can be drawn into competing meaning systems and modes of identification.

Nevertheless, despite some recognition of continuities between structuralism and post-structuralism, in concluding this section it is worth reiterating the differences between them. The main distinction between the two approaches is that between 'depth,' and 'breadth', structuralism sought, in a formalistic and scientistic manner, to identify the 'deep' structures that determine 'surface' behaviour; post-structuralism regards the search for such deep structures as misplaced. 'Surface' behaviour arises from relations that are also to be found on the surface (that is the relation between the reader and text): thus a much greater plurality of meanings, behaviours and identities is revealed by post-structuralism and this plurality lies at the centre of its deliberations. These distinctions are usefully summarized by Gibson-Graham.

> The philosophers who were to become known as 'post-structualists' confronted the structuralist project with a sceptical attitude toward determination by 'underlying' structures and attempts to grasp the ultimate 'truth' of language, culture, society and psyche. But perhaps their most salient move was to call into question the fixed relationship between signifier and signified that characterised Sausserian linguistics. From a post-structuralist perspective, language does not exist as a system of differences among a fixed set of signs. Rather the signifier–signified relations that generate meaning are continually being created and revised as words are recontextualised in the endless production of texts. The creation of meanings is an unfinished process, a site of (political) struggle where alternative meanings are generated and only temporarily fixed. Thus the meaning of the word 'woman' in the context of 'husband', 'work', and 'politics'. Political struggles undertaken by feminists can be seen as multiplying the contextualisation and significations of 'woman' and, in the process, destabilising the fixities of meaning associated with a particular order. (2000: 96)

Post-structuralism, then, describes social and cultural systems that are open and dynamic, constantly in the process of 'becoming'. The task of post-structuralist theorizing is to trace the resulting trajectories of change. In doing so, it must not only look back across the 'line of flight' (as the post-structuralist philosopher Gilles Deleuze would put it, see Chapter 4 below) but must also attempt to situate itself on the cusp of motive forces so the full extent of (post-) structural transformation can be apprehended.

Post-structuralism and geography

It is fair to say that structuralism of the Saussurian and Lévi-Straussian variety had only limited impact upon the conduct of human geography. Perhaps it was the focus on texts, kinship systems and mythologies in the structuralist literature, or the overriding concern with 'deep', all-determining structures, that rendered the approach unpalatable to mainstream geographers. Whatever the reason, the structuralist movement found few geographical adherents.[2] Nevertheless, some work of a recognizably 'structuralist' kind has been conducted within geography. One example is the 'spatial science' that emerged during the 1950s and 1960s. This work sought to simplify geographical variety into a few (measurable) factors – such as distance and efficiency – which would function to explain location practices and movement patterns (see Haggett, 1965, for an overview). It thus employed a spatial 'code' to interpret geographical form. But perhaps a better illustration is the structuralist Marxism that became popular amongst radical geographers during the 1960s and 1970s. This work drew heavily upon the writings of Louis Althusser, a Marxist theorist who was profoundly influenced by structuralism (Gutting, 2001). In Althusser's view, Marx provided a *scientific* perspective on societal development that could best be appreciated from within the structuralist framework. In keeping with mainstream structuralism, structuralist Marxism would look beyond the actions of individuals and social movements (operating in a context of class conflict) to the determining structures that lie 'beneath' any social formation. At the same time, the structuralist concern for structural patterns would be recast in terms of the Marxist distinction between a 'base' of economic or productive forces and a 'superstructure' of political, cultural and social formations. In Althusser's view, all these structures interact with one another so that each can make a contribution to the shape of any given society. However, he sought to retain Marx's materialism by arguing that the economic will inevitably be deterministic 'in the last instance'. Thus, the structuralist focus on structural patterns re-emerges in the form of non-economic structures arranged in layers above the economic base. Moreover, the non-economic structures retain only a relative autonomy from this base; thus, any 'deep' explanation of society inevitably leads to an *economic* account of social change.

Althusserian Marxism proved influential in human geography, with geographical work on pre-capitalist societies, the state, urban form, and spatial divisions of labour all showing evidence of Althusserian influences (see Peet and Thrift, 1989, for an overview). While this approach was superseded during the 1980s and

1990s by a form of Marxism that owed more to realism than to structuralism, commonalities between the two were evident. For instance, realism, like Althusserianism, sought to delve beneath the surface of society to the 'causal powers' of particular social structures, with the economic again retaining its pre-eminence. Realist Marxism aimed to combine abstract analysis of basic structures and mechanisms with empirical studies of 'concrete' outcomes (see Sayer, 1984).

In Law and Urry's (2004: 397–8) view, Marxism of this structuralist or realist variety tends to produce a geography of highly-structured social spaces. As they say, 'Marxist theory mobilises a range of metaphors, but a notion of levels is carried in many of them – as in the distinction between the causal 'in the last instance' infrastructure and the 'caused' superstructure'. This theorizing of 'levels', they suggest, gives rise to a 'Euclidean' spatiality associated with height, depth, size and proximity. Thus, structuralist theory sees space as a surface configured by the play of underlying structures: 'Using metaphors that are more or less Euclidean this means that [structuralist theory] tends to enact and produce a Euclidean reality of discrete entities of different sizes contained within discrete and very often homogeneous social spaces' (Law and Urry, 2004: 398). The geography produced by structuralism, then, is a geography of well-ordered, topographical spaces.

BOX 1.3

The terms 'topography' and 'topology' will be used extensively in the following chapters. Thus, it is useful to clarify their meanings:

- 'Topography' is defined by *Cassell's Concise English Dictionary* as: 'the detailed description of particular places; representation of local features on maps, etc.; the artificial features of a place or district; the mapping of the surface or the anatomy of particular regions of the body'. This definition roughly approximates the use of the term in post-structuralist geography. Topographical spaces are seen as 'contained spaces', in which space is seen in terms of its surface (maps, points, lines, contours and so forth). Topographical space is also sometimes called 'Euclidean' space because of this concern for contained surfaces.
- 'Topology' is defined by *Cassell's Concise English Dictionary* as 'the study of geometrical properties and relationships which are not affected by distortion of a figure'. The term 'topology' is not a standard geographical concept and has been adapted by post-structuralist writers from mathematics. It refers not to surfaces but to 'relations' and to the interactions between relations. It therefore enables geographers to go below the surface to study processes of spatial emergence. It suggests that any spatial coherence that is achieved (on the surface) serves to disguise the relational complexities that lie 'underneath' spatial forms.

It was against this background that post-structuralism made its incursions into the geographical domain. Initially, its impact comprised a shift in geographical attention away from the economy and towards 'culture'. As we noted above, post-structuralism emerged from the analysis of texts and the generation of textual meanings. From there it travelled easily into the cultural domain and showed how broad swathes of culture could be 'read' in a textual fashion (see Smith, 2001). The emergence of post-structuralism in geography coincided with the rise of cultural geography as a part of the geographical mainstream. Now geographers could use textual analysis to 'read' geographical cultures (for example, landscapes – see Cosgrove and Daniels, 1988). These 'readings' could open up fresh perspectives on taken-for-granted objects of analysis and could allow an engagement with plural and multiple forms of identity. As Derek Gregory (1994: 75) summarizes it: 'the closures and certainities of the objectivist tradition within human geography [had become] increasingly suspect [...] a kind of strategic reversal ha[d] been put into effect, which now continually unsettle[d] attempts to claim a synoptic completeness for the geographical project'. In short, post-structuralism allowed new theoretical openings to be made; it enabled the creation of new spatial imaginaries, which seemed to stem from outside the closed worlds of spatial science and structuralist Marxism. Thus, it was argued that 'any critical human geography must attend to the ways in which meanings are spun around the topoi of different lifeworlds, threaded into social practices and woven into relations of power' (Gregory, 1994: 76).

For some geographers this meant seeking out spaces that had been neglected by earlier geographical traditions. The aim now was not only to study such spaces but also to empower and enfranchise ('give voice' to) social groups that had been neglected by conventional geographical approaches (for example, Jackson, 1989; Keith and Pile, 1993; Rose, 1993; Bell and Valentine, 1995). For others, it meant focusing squarely on the forces that were systematically disempowering and disenfranchising marginalized social groupings (Castells, 1983; Harvey, 1989; Cresswell, 1996; Soja, 1996). Inevitably, these two sets of concerns – the investigation of the experience of being marginalized and greater understanding of the processes of marginalization – were brought together so that geographies of 'resistance' came to the fore, giving renewed emphasis to the post-structuralist insight that meaning and identity arise from an interaction between system-wide relations and divergent 'readings' of those relations (cf. Pile and Keith, 1997). The new geographies of resistance aimed to highlight how social groups and social actors work to subvert and appropriate space in the face of hegemonic tendencies within the system as a whole. Thus, much effort was expended on showing the widespread nature of resistance across diverse spatial locations (an effort that ultimately led some commentators, such as Cresswell (2000) to conclude that almost any social practice of any kind could ultimately be given the label 'resistance').

As rich descriptions of diverse groupings and their spaces and places were provided, so geography came to engage with multiple perspectives, multiple

spaces and multiple sets of (spatial) relations. As Massey (1991: 28) notes: 'If it
is now recognised that people have multiple identities then the same point can
be made in relation to places'. For many geographers, the advance of feminism
(Bondi, 1990), post-colonialist studies (Corbridge, 1993) and queer theory
(Binnie, 1997) opened up the new spaces of enquiry. These spaces could all
be seen as 'disruptions' of, or 'commotions' in, the spatial orders that had been
established by earlier geographical approaches (such as spatial science, with
its distances, lines and surfaces, and Marxism, with its carefully layered social
strata). They therefore helped to undermine taken-for-granted notions of robust
and enduring spatial structures, forever imposing strict patterns of spatial ordering throughout given societies.

For Ed Soja (1996) these 'marginal' spaces all come together in a zone he
calls 'thirdspace'. This term is added to the categories of 'firstspace' (the formal
arrangement of things in space) and 'secondspace' (representations and conceptions of space) that Soja finds in Lefebvre's (1991) work. Soja describes
'thirdspace' in the following way:

> Thirdspace [...] is portrayed as multi-sided and contradictory, oppressive and liberating,
> passionate and routine, knowable and unknowable. It is a space of radical openness, a site
> of resistance and struggle, a space of multiplicitous representations, investigatable through
> its binarized oppositions but also where *il ya toujours l'Autre*, where there are always
> 'other' spaces, heterotopologies, paradoxical geographies waiting to be explored. It is a
> meeting ground, a site of hybridity and *mestizaje* and moving beyond entrenched bound-
> aries, a margin or edge where ties can be severed and also where new ties can be forged.
> It can be mapped but never captured in conventional cartographies; it can be creatively
> imagined but obtains meaning only when practiced and fully lived. (1999: 276)

This passage neatly captures the post-structuralist influence on geographies of
resistance and marginality and indicates how multiplicity has become a central
aspect of geographical inquiry. It displays an acute sensitivity to the openness
of space and the importance of new ways of being in space. However, it also
raises one or two issues that have come to worry proponents of the post-
structuralist approach. These worries are elaborated by Mitch Rose (2002) in
his review of 'resistance studies' within human geography. To begin with, Rose
acknowledges the debt that work on strategies of spatialized resistance owes to
post-structuralism: 'since the introduction of new cultural geography, and its
infusion of literary theory and cultural studies, social systems have been con-
ceptualised as symbolic contexts made "partially" stable by hegemonic relations
of power' (Rose, 2002: 384). However, he then goes on to question the uses
that have been made of post-structuralist theory in the analysis of resistance
strategies: 'although cultural geographers recognise the forces of contestation
and change circulating within systems, they nonetheless conceptualise the
system itself as something originally stable upon which deconstructive forces
act' (2002: 384). Rose's criticism here seems to be that residues of structuralism remain salient within post-structuralist geography. In his view, 'resistance

studies conceptualise agents as responding to a dominant system [...] in doing so [they] constitute the structural nature of the system as primary'. In other words, geographically positioned social actors are reacting to an already stabilized system, one that seems detached from, and alien to, their own social practices. In contrast, Rose argues that social categories 'are never stabilised, normalised, sedimented or structured'. Rather, 'they are always in a process of dynamic unfolding and becoming' (2002: 385).

Rose here echoes criticisms made slightly earlier by Doreen Massey (2000: 280) of de Certeau's (1984) work on practices of resistance in the contemporary city. Massey argues that de Certeau uses a model of power which essentially 'opposes the strategies of the powerful to the tactics of the little people, those who resist'. She then goes on to say that this model is 'an engagement with structuralism – an attempt to find a way out of a system which in fact has no exit signs. What de Certeau does is to retain 'the structure' as his conceptual starting point, and then recruit guerrillas to attack it'. Like Rose, Massey argues for a jettisoning of geographical post-structuralism's residual structuralism. In so doing, she suggests that geographers need to take a 'performative' approach to the study of spatially contextualized resistance strategies. In making this proposal, Massey leads us towards another strand of post-structuralism in human geography, one that carries some rather profound implications for geographical thinking on the nature of space.

The need for a 'deeper' reconceptualization of space within geographical theory has been made especially forcefully by Nigel Thrift in a series of stimulating and provocative publications (for example, 1996, 1999, 2004a, 2004b). Following post-structuralist work on spatial practice, Thrift argues that geography should move away from a sense of space 'as a practico-inert container of action' and should now begin to conceptualize space as a 'socially produced set of manifolds' (Crang and Thrift, 2000: 2). Various reasons are given for this shift:

1. Geographers now recognize that the human subject is not just implicated in meaningful action but is also implicated in *embodied* action. Thus, humans act within 'spaces of embodiment' and react to other embodied entities.
2. The embodied subject finds itself in an *object* world. Here 'thought itself comes heavily equipped, surrounded by a vast apparatus of devices and metrics which are not incidental but through a series of mediated shifts produce their own object' (Crang and Thrift, 2000: 19).
3. The actor's contextualization in an embodied object world means that action is distributed across heterogeneous relations with the result that space can no longer be seen as a 'container' but must be seen as an active presence in social practice.

These observations lead Thrift into developing what he calls 'non-representational theory' (Thrift, 1996). Put simply, non-representational theory is based on an acceptance of the view that 'we cannot extract a representation of the world because we are slap bang in the middle of it, co-constructing it with numerous human and non-human others for numerous ends' (Thrift, 1999: 296–7). In other words, non-representational theory acknowledges that there are 'very strong limits on what can be known and how we can know' (1999: 303); thus, 'the varieties of stability we call "representation" [...] can only cover so much of the world' (Thrift, 2004a: 89). Limits to representation derive ultimately from 'our' embodiment in space (and time). Following recognition of embodiment, and the multiplicity of relations that connect humans both to given spaces and to other humans, we need to admit that there are 'numerous perspectives on, and metaphors of, what counts as knowledge, or more accurately knowledges' (1999: 303). Thrift therefore follows Donna Haraway (1991) in arguing that knowledge is always 'situated'. The situatedness of knowledge carries certain important implications for geographical theory:

1. Concepts must be seen as 'indefinite' – that is, they are open and fluid, their main purpose is not to 'represent' but to 'resonate' (Thrift, 1999: 304).
2. Knowledge is always contextualized, it is always located in space, notably in an embodied, material space. Moreover, context is 'performative', it is 'a plural event which is more or less spatially extensive and more or less temporally specific' (Thrift, 1996: 41).
3. Theory is not oriented to the apprehension of (a single) truth but is 'a practical means of going on' (Thrift 1996: 304); it is a way of engaging with the world that recognizes its own contextual limitations (in this sense, it encourages the theorist to engage in 'reflexivity' – that is, a reflection on his or her situatedeness).
4. Non-representational theory promotes 'relational rather than representational understandings' (Thrift, 1996: 304) because embodied subjects are necessarily involved in multiple encounters and interactions. The theory thus emphasizes the 'flow of practice in everyday life' and the 'on-going creation of effects through encounters' rather than 'consciously planned codings and symbols'. The 'everyday is therefore seen as a set of skills which are highly performative' (Thrift and Dewsbury, 2000: 415).

This new concern for a(n) (embodied) subject, tightly woven into the fabric of space (-time), leads Thrift to endorse those varieties of post-structuralism that acknowledge most clearly the importance of 'physical presences and absences (as well as linguistic presences and absences)' (1996: 31). As a result, he claims that the works of Foucault and Deleuze (especially the latter) are more helpful to human geographers than the works of Derrida and Lyotard (Thrift, 1996). On this view, Foucault and Deleuze make an important move 'beyond' the text to engage with embodied practices and object worlds (Thrift, 2004b). Moreover, in their work, social practice is not seen as 'localised in individuals but is understood as a relational structure' (Thrift, 2004a: 87). Relationality here carries important implications for the analysis of space and place because, rather than seeing either of these geographical phenomena as fixed and contained, Foucaultian and Deleuzian versions of post-structuralism conceptualize space and place as 'territories of becoming that produce new potentials' (Thrift, 2004a: 88). Such potentials derive from the openness of space and place, from the way social relations and spatial relations intersect and combine: space is practised and performed in the same way that social identity and belonging are practised and performed. In sum, then,

> non-representational theory takes the world to be a kaleidoscopic mix of space-times, constantly being built up and torn down. These space-times normally co-exist, folding into one another, existing in the interstices between each other, creating all manner of bizarre and unexpected combinations [...] Some space-times are more durable. Their reach is able to be extended by intermediaries, metrics and associational knowledges [...] Other space-times flicker out of existence. (Thrift, 2004a: 91)

Here, we encounter a dynamic, almost visceral form of post-structuralist thought, one that engages with the turbulent nature of space-time and the social processes and practices that lead some spaces to endure while others 'flicker out of existence'. As Latham (2003: 1902) puts it, this is a geography of the 'event', one that engages with the complex entanglements of social prac-tice and the fleshy materialities of the socio-spatial world. It situates itself on the cusp of trajectories of movement and seeks to identify how these trajecto-ries unfold over time and through space.

The 'geography of becoming' that animates non-representational theory has probably been taken furthest in the writings of Marcus Doel (1999, 2000, 2004). In Doel's view, geography has long been beset by the problem of 'pointillism': as he puts it, 'in geography the fundamental illusion is the auton-omy and primacy of the point' (Doel, 1999: 32). This illusion, Doel argues, has dominated geographical practice. Thus, academic labour in geography typically involves three interrelated tasks:

1. *enumerating* the properties and attributes of various spatial entities ('a com-prehensive and encyclopaedic task of ascription and description, which lends itself immediately to empiricism and story telling' – Doel, 1999: 120);
2. *mapping* surface phenomena into regular units of classification ('a carto-graphic and diagrammatic task of relative localisation, which thereby assumes a frame of reference and orientation' – Doel, 1999: 120);
3. *synthesizing* a variety of socio-spatial processes in the context of circum-scribed geographical locations ('an integrative and unifying task of con-textualisation, actualisation, and localisation that more often than not leads to their paralysis and decomposition' – Doel, 1999: 121).

In Doel's account, these various tasks give human geography a 'superficial' quality in that it 'clings to the surface of what actually takes place' (Doel, 1999: 121–2). Thus,

Everywhere one looks, geography and geographers are hung up on points: sites, places, nodes, integers, integrands, wholes, digits, identities, differences, the self, the same, the other, positions, op-positions, bifeds, trifeds, and so on and so forth. Lines run between points. Surfaces are extended from lines. Volumes are unfolded from surfaces. And then there is the networking, not to mention the hybridisation, othering, thirding [...] Etcetera. In sum, spatial scientists have suspended themselves between all manner of points, and that is their undoing. (Doel, 2000: 120)

What Doel seeks is an engagement not with points and lines but with 'the permanency of essential forms' (Doel, 1999: 123). Thus, it would be better if geographers could approach space 'as a verb rather than a noun. *To space* – that's all. Spacing is an action, an event, a way of being [...] Space is immanent. It has only itself' (Doel, 2000: 125). In short, then, for Doel (1999: 6) 'post-structuralist geography affirms what is still coming'.

BOX 1.4

Some core features of post-structuralist geography:

- Spaces and places should not be seen as closed and contained but as open and engaged with other spaces and places. These engagements mean that spaces and places are cross-cut by differing processes and practices, some that emanate from within, some that emanate from without.
- Spaces and places are therefore multiplicities – that is, they are made of differing spatial practices, identifications and forms of belonging.
- There can be acute struggles over whose 'reading' of space should take priority. Thus, strategies of domination and resistance ensue around spatial identities and spatial practices.
- The outcome of these struggles must be seen as under-determined by existing spatial structures. Rather, struggles can lead to the need for spatial 'openings', new forms of spatial identity and new forms of spatial practice.
- The 'performance' of social practice and the performance of space go hand in hand. Space is therefore not fixed but mutable.
- Moreover, the notion that the 'performer' (for example, the social agent) and the context of performance (for example, the space or place) are distinct from one another should be abandoned: both are entangled in the heterogeneous processes of spatial 'becoming'.

The post-structuralist geography of Thrift and Doel takes us beyond simply an engagement with new objects of analysis or broader landscapes of geographical concern (that is, Soja's 'thirdspace'); rather, it engages with the very 'stuff' of space, the fleshy materialities that make up both human bodies and spatial textures. It embraces the complex entanglements that inevitably link the social and the spatial and shows how geography must apply at precisely the point of 'crossover' between the social and the material. In this regard, Doel and Thrift's post-structuralism accords nicely (though perhaps unexpectedly) with Peet's description of geography given at the beginning of this chapter. Rather than separating the natural and the social into two discrete branches of geographical inquiry ('human geography' and 'physical geography') this type of post-structuralist geography celebrates the complex interactions that take place between the two 'domains'.[3] Thus, 'biological' or 'ecological' post-structuralism (Massumi, 2002; Thrift, 2004b) potentially reinvigorates geography and demarcates more clearly its distinctive characteristics. No longer need geography render the world in textual form (in the style of 'cultural studies'); now it can focus squarely on its own core concern – nature–society relations, and the 'fleshy materialities' that emerge as the two domains combine.

Geography and relationality

> [W]e will only unlock the power of post-structuralist geography to the extent that we embrace nothing but relations and co-relations, their folding and unfolding.
> (Doel, 2004: 147)

The study of spatial relations has long constituted a key component of geographical work. David Harvey (1996), for instance, traces a relational lineage from the work of Leibniz and Whitehead through to contemporary studies of social justice and social nature. From Leibniz, Harvey takes the idea that space is not a 'container' but is something that is always dependent on the processes or substances that go into 'making it up'. From Whitehead he derives the insight that these processes and substances are constituted from relations. Thus, any kind of spatial 'permanence' arises as 'a system of "extensive connection" out of processes' (Harvey, 1996: 261). The process of place formation then becomes 'a process of carving out "permanences" from the flow of processes creating spaces. But the "permanences" – no matter how solid they may seem – are not eternal: they are always subject to time as "perpetual perishing". They are contingent on the processes that create, sustain and dissolve them' (Harvey, 1996: 261). In this view, space is made not by (underlying) structures but by diverse (physical, biological, social, cultural) processes; in turn, these processes are made by the relations established between entities of various kinds. As Harvey (1996: 294) discusses, this relational perspective leads to us into seeing discrete spaces and places as 'dynamic configurations of relative "permanences" within the overall spatio-temporal dynamics of ecological processes'. We shall return to the ecological aspect of this in later chapters (notably Chapter 8), but for now we need to go a little further into the notion of 'relational space' in order to show how it links to post-structuralist materialities.

In Harvey's account the general character of space is given by the processes that somehow stabilize (semi-) permanent spatial assemblages. Thus, space is generated by interactions and interrelations. Human geographers, then, need to account for the relational spaces that *do* emerge and they need to understand how particular spatial configurations are generated. But equally, some attention must be paid to spaces that *do not* emerge, to the sets of relations that fail to gain any kind of spatial coherence. Relations *between* relations therefore become important. The shape of space can be seen as the 'expression' of 'underlying' relations; but it can also be seen as the suppression of all those other relations that might have gained some amount of permanence had they not 'flickered out of existence' in Thrift's (2004a: 91) telling phrase. The relational making of space is both a consensual and contested process. 'Consensual' because relations are usually made out of agreements or alignments between two or more entities; 'contested' because the construction of one set of relations may involve both the exclusion of some entities (and their relations) as well as the forcible enrolment of others. In short, relational space is a 'power-filled' space in which some alignments come to dominate, at least for a period of time, while others come to be dominate*d*. So while multiple sets of relations may well co-exist, there is likely to be some competition between these relations over the composition of particular spaces and places.

These various aspects of relational space have been explored in detail by Doreen Massey (1992, 1998, 1999b, 2005). Like Harvey, Massey wishes to move away from a structuralist conception of space. In fact, she believes that structuralist theory has had great difficulty in accounting for the significance of space: in her view it tends to see the spatial realm simply in terms of partitions and closures. In contrast, Massey proposes a relational approach. She (1998: 27–8) helpfully outlines three basic propositions that she sees as intrinsic to the approach:

1. Space is a product of interrelations, as identified above. These interrelations run through differing spatial scales from the very local to the global and all points inbetween.
2. Space is the sphere of the possibility of multiplicity. Because various relations 'run through' space – that is, compose space – all may come into being *spatially*. As Massey (1998: 28) puts it, 'without space, no multiplicity; without multiplicity, no space [...] Multiplicity and space are co-constitutive'.
3. Space is never closed, never fixed. In other words, space is always in the process of becoming as relations unfold: 'there are always – at any moment "in time" – connections yet to be made, juxtapositions yet to flower into interaction (or not, for not all potential connections have to be established), relations which may or may not be accomplisheded' (Massey, 1998: 28).

These three points reinforce Harvey's argument that spaces are (provisionally) stabilized out of complex, open-ended processes. Massey (1998: 29) also believes that the three points link to core facets of post-structuralism:

1. First, the notion that space is made from relations chimes well with post-structuralism's anti-foundationalist and anti-essentialist view of politics: 'that is, in place of a kind of identity politics which [...] argues for the

rights of, or claims to equality for [...] already constituted identities, this anti-essentialist politics takes the constitution of the identities themselves to be one of the central stakes of the political'. Moreover, 'identities/entities, the relations "between" them, and the spatiality which is part of them are all co-constitutive' (Massey, 1998: 29).

2. The idea that space is the sphere of possibility of multiplicity accords with the long-standing post-structuralist concern for 'difference'. Yet, we must now accept that differences are always spatialized, always positioned in space.

3. As space is a process of becoming, it is always in the process of being made and is always (likely to be) unfinished: 'there are always loose ends in space' (Massey, 1998: 37). Moreover, because space is made from competing and co-existing relations, it holds an unpredictable character that can potentially generate 'new spaces, new identities, new relations and differences' (Massey, 1998: 38). Openness and newness thus go hand in hand.

Space becomes, in Massey's (1991) terminology, 'a meeting place'; it is where relations interweave and intersect. In 'meeting places', relational conflicts can emerge just as consensual relations can be consolidated. Importantly, the relations that run through meeting places run over differing spatial scales. As Massey (1998: 37) puts it: 'space [...] is the product of the intricacies and complexities, the inter-twinings and the non-interlockings, of relations, from the unimaginably cosmic to the intimately tiny'. Thus, this relational perspective encourages us to rethink the meaning of spatial scale and the way relations are invariably consolidated between scales. In particular, as Ash Amin (2002: 391) notes, it appears to suggest that scale might be conceptualized in 'non-territorial terms'. By this, he means that differing spaces and places can be seen not as hierarchies (global, national, local) but as 'nodes in relational settings'. Thus, scale becomes distance, or, more accurately, the 'length of relation'. Places are bound to one another relationally: the significance and composition of the relations defines the significance of scale.

BOX 1.5

Some significant features of relational space:

*   Space is not a 'container' for entities and processes; rather space is made by entities and processes. Moreover, these entities and processes combine in relations. Thus, space is made by relations. Space is relational.
*   Discrete spaces and places are stabilizations of processes and relations. In David Harvey's terms, they are 'permanences'. However, these 'permanences' are not

Continued

permanent for they are only stabilized provisionally. They must be continually remade and as they are remade so they change.

- Space is made of multiple relations. These relations meet in space, at meeting places. There can be conflicts as sets of relations jostle for spatial supremacy. Equally, there can be consensus as alliances are built and alignments are forged.
- Spaces are open not closed. As multiple relations meet in space so new relations are formed and new (spatial) identities come into being. The openness of space also means that spaces and places are dynamic rather than static. In other words, they are always in the process of becoming. Geography must trace the trajectory of change and the line of force.

There are, then, no essential qualities to any given place (it is a 'global' place, a 'local' place), for all (scalar) identities are derived from the relations established between places. This should not be taken to imply, however, that multi-scalar relations have no territorial effects: as Massey explains, a 'power-geometry' immediately emerges once relations meet in space:

> different social groups and individuals are placed in very distinct ways in relation to these flows and interconnections. This point concerns not merely the issue of who moves and who doesn't, although that is an important element of it; it is also about power in relation to the flows and the movement. Different social groups have distinct relationships to this anyway differentiated mobility: some people are more in charge of it than others; some initiate flows and movement, others don't; some are more on the receiving end of it than others; some are effectively imprisoned by it. (1991: 25–6)

We can see here that Massey skilfully ties together the two post-structuralist strands identified above. On the one hand, she uses the notion of relational space to show how multiple processes combine to bring particular spatial formations into being. She suggests that what gives any place its specificity (or, following Harvey, its 'permanence') is the constellation of relations that meet and weave together at a particular locus ('if one moves in from the satellite towards the globe, holding all those networks of social relations and movements and communications in one's head, then each 'place' can be seen as a particular, unique, point of their intersection [...] a meeting place' – Massey, 1991: 28). On the other hand, Massey expresses concern for social groupings that may find themselves marginalized by dominant relational configurations (see also Amin and Graham, 1997). As she notes, the means by which people are 'placed' within given sets of relations can either strengthen or weaken their ability to exercise some degree of control over those very relations. In other words, just because space is 'relational' does not mean it is less restricting or confining. Relations are inevitably double-edged: they can facilitate movement and access; equally they can entrench confinement and exclusion. Thus, spatial relations are also power relations. As we shall see in subsequent chapters, the study of confinement and movement as expressions of spatialized power relations is a key concern of post-structuralist human geography.

Conclusion

In this chapter we have begun to consider the character of post-structuralist geography. We have seen how post-structuralism emerged from within the structuralist movement during the 1960s and 1970s. We have seen how structuralism was reinvented as post-structuralism and how this reinvention entailed a shift in theoretical focus from closed and deterministic structures to open and dynamic relations. To begin with, post-structuralism directed its attention to the relationship between the reader and the text. It argued that texts contain multiple narratives and that these can be opened up by attentive and active readers. The concern for meaning generation in texts quickly spread to a concern for meaning generation in other areas of social and cultural life. Thus, post-structuralists came to study sources of identity and the way multiple forms of identity flow from the complex systems that surround social actors.

This general form of post-structuralism soon found its way into geography. To begin with, post-structuralist geographers argued that there are many more spaces than those to be found in standard geographical textbooks. Moreover, they asserted that these 'alternative' spaces are closely tied to 'alternative' forms of identity. Thus, meaning, identity and space became closely intertwined. As geographers searched for 'alternative' modes of spatialization, so they came to focus on the complex sets of relations that inevitably surround any spatial entity. Many of these relations were conflictual and oppositional; thus, geographies of dominance and resistance emerged with the consequence that the making of space by either dominant or marginal groups came to be seen also as an exercise in power relations. Post-structuralist geography quickly became a geography of power and its various spatial entanglements.

At the same time, other geographers employed post-structuralist theory to ask some basic and profound questions about the composition of space itself. This led some to argue that space is much more complex and dynamic than many spatial analysts appear to realize. Moreover, spaces are made of complex sets of relations so that any spatial 'solidity' must be seen as an accomplishment, something that has to be achieved in the face of flux and instability. Space is made and it is made relationally. This means that space and place have no determining structure; rather, structure is an effect of relations. Moreover, spatial relations reach across spatial scales, indicating that geographical scale is also an outcome of relational processes and actions.

Summarized in this way, it appears that post-structuralism has had a profound impact on the discipline of geography. However, it should be noted that post-structuralist geography is still to be found some distance away from 'mainstream' geography (although it is undoubtedly a lot closer to the centre of the discipline than it once was). Moreover, many of the arguments put forward by post-structuralist geographers have met with considerable criticism. In fact, there are still crucial debates running through geography about the nature of the post-structuralist challenge and the implications it holds for the discipline as a whole (for example, Martin, 2001; Hamnett, 2003).

Nevertheless, the above account indicates that post-structuralism has raised many significant issues for human geographers and has led to important new theoretical challenges. In the following chapters, these challenges will be studied in some detail. However, to return to points made in the introduction to this chapter, only a partial and selective coverage of post-structuralism will be provided. In part, this is because the literature that falls under the heading of 'post-structuralist geography' is too vast to easily survey in one text (it should be noted that only a carefully chosen selection has been reviewed in the preceding pages). But the main reason for taking a discriminating route through the thickets of post-structuralist theory is in order to concentrate on those approaches that hold most promise for the study of nature–society interrelations (to return once more to the Peet quote provided at the beginning of the chapter). Thus, in the next few chapters we will look at the work of theorists that fit readily into the category of 'non-representational theory' (as outlined by Thrift, 1996). We will examine forms of post-structuralism that are situated 'beyond the text' in the 'fleshy materialities' of the bio-social domain. In the main, the analysis focuses upon the work of Michel Foucault, Bruno Latour, John Law, Gilles Deleuze and Michel Serres. While there are many differences between these various thinkers (as will become evident below), they all engage to some extent with heterogeneity and relationality. By considering their geographical contributions we begin to get a feel for the full significance of relational thinking in post-structuralist geography and the way such thinking might be employed in the study of differing spatial formations.

SUMMARY

This chapter has shown how 'post-structuralism' emerged from 'structuralism' and also how post-structuralism has affected geography. It suggested that post-structuralism's influence manifests itself in two main ways: first, it leads to a concern for spaces of multiplicity; second, it challenges some basic geographical assumptions about the make-up of space itself. In particular, it proposes that space is made not of structures but of relations. Thus, a new geography of spatial relations has emerged.

FURTHER READING

For an excellent introduction to post-structuralist theory, see Gary Gutting's (2001) *French Philosophy in the Twentieth Century* and Catherine Belsey's (2002) *Poststructuralism: A Very Short Introduction.* An overview of post-structuralism in human geography can be gained from Richard Peet's (1998) *Modern Geographical Thought.* A more challenging account can be found in Marcus Doel's (1999) *Poststructuralist Geographies.* On 'non-representational theory', see Chapter 1 of Nigel Thrift's (1996) *Spatial Formations.* For a comprehensive overview of relationalism, see Doreen Massey's (2005) book, *For Space.*

Notes

1. For instance, Cloke et al. (1991: 196–7), in a general introductory text, say that scientific or narrative writings 'risk imposing order and indifference upon the subject matter being addressed'. Thus, 'we need to attend very seriously to the discipline's "textual strategies" if we are ever to capture the differences, complexities, nuances, achievements and sufferings that are the postmodern geographies of the contemporary world'. The term 'post-structuralism' could easily be exchanged for 'postmodernism' in this passage.
2. One exception might be the historical geographies of Ferdinand Braudel. Along with other members of the so-called *Annales* School, Braudel produced histories rooted not in the actions of individuals or social movements but in geography, climate, terrain and natural resources. However, while Braudel's histories carried many implications for geography, they were primarily set within the discipline of history and will therefore be disregarded here. For an accessible introduction, see Braudel (1977).
3. As Thrift (2004b: 59) notes, 'distance from biology is no longer seen as a prime marker of social and cultural theory [...] It has become increasingly evident that the biological constitution of being (so-called "biolayering") has to be taken into account if performative force is ever to be understood'.

Part 1 Theories

Introduction

One objective of the analysis presented in the following pages is to trace theoretical continuities through differing versions of post-structuralism. With this objective in mind, the theoretical part of the book begins by assessing the contribution of Michel Foucault to both the development of post-structuralist geography and understandings of relational space. Foucault provides a useful starting point because his work spans the structuralist/post-structuralist shift identified in Chapter 1. In his early writings, Foucault still operates within the structuralist paradigm, and while he has some interesting things to say about space during this phase, he only becomes a significant geographical thinker once he moves more fully into (what we now call) post-structuralism. Foucault's later writings provide us with some wonderfully detailed and insightful geographies of relational space, notably in the context of his studies of discipline and government. Foucault's 'geography' will therefore be assessed and its strengths and weaknesses will be evaluated.

One problem that arises in Foucault's work is the difficulty of moving convincingly between spatial scales. In seeking to counter this weakness, we examine the work of Bruno Latour, a scholar who follows broadly in Foucault's footsteps but who allows us to more fully appreciate the relational nature of spatial scale. Latour traces the way relations are 'made' through space and identifies how resources of various kinds are utilized in the process of relationship building. In short, he provides us with a geography of heterogeneous associations (Murdoch, 1997). This geography is revealed most clearly in his studies of scientific practice. These studies reveal how the conduct of science requires the establishment of complex networks that run between scientific 'centres' and non-scientific 'peripheries'. In illustrating how scientific facts and artefacts move through these networks and out into the world, Latour shows how the 'local' and the 'global' emerge as network effects. On this account, spatial scales are not stacked on top of one another in discrete layers; rather, scale is generated by distance – that is, it stems from the consolidation of power relations between dispersed sites. For Latour, there is no macro- or micro-level of social reality; there are just sets of relations, some long, some short.

We then move on to consider how networks make space. Drawing upon John Law's studies of differing network spatialities, we see that spatial relations can be either strongly prescriptive or relatively fluid in nature. In other words, space can be strongly configured by 'powerful' networks or can be made up by many competing sets of associations. In this latter case, space may remain fluid and open rather than singular and closed. Given these varied outcomes, it is suggested we need to examine in detail how differing spaces are made by differing sets of relations. This takes us into the realm of multiple spaces and multiple relations. Thus, a concern for 'multiplicity' comes to the fore.

The close entwining of networks and space in Law's analysis gives rise to a key question: can geography still provide a general overview of the spatial domain or must it be confined to specific network perspectives? This question is addressed through the work of two well-known 'geo-philosophers', Gilles Deleuze and Michel Serres. From these two thinkers, we gain an appreciation of space as multiple, striated and undulating (in line with relational processes). In describing landscapes of singularity and multiplicity, Deleuze and Serres help us to link the post-structuralist interest in relationality to the geographical concern for territory. They show that processes of (network) emergence necessarily co-exist with zones of (network) stability. In short, they alert us to the simultaneous existence of topography and topology and suggest ways in which these two spatial forms might be related to one another within post-structuralist geography.

2

Spaces of discipline and government

Foucault is one of many who want a new conception of how power and knowledge interact. But he is not looking for a relation between two givens, power and knowledge. As always, he is trying to rethink the entire subject matter, and his knowledge and power are to be something else. Nobody knows this knowledge; no one wields this power. Yes, there are people who know this and that. Yes, there are individuals and organisations that rule other people. Yes, there are suppressions and repressions that come from authority. Yes, the forms of knowledge and of power since the nineteenth century have served the bourgeoisie above all others [...] But those ruling classes don't know how they do it, nor could they do it without the other terms in the power relation – the functionaries, the governed, the repressed, the exiled – each willingly or unwillingly doing its bit. One ought to begin an analysis of power from the ground up, at the level of tiny local events where battles are unwittingly enacted by players who don't know what they are doing. (Hacking, 2002)

Introduction

There is little doubt that Michel Foucault would have resented his inclusion in a book on 'post-structuralism'. For most of his intellectual life Foucault appeared determined to escape any such crude intellectual classification – he once said his historical studies comprised an attempt to escape from any fixed identity, 'to have no face' (Foucault, 1972: 17). Moreover, Foucault disavowed any theoretical or methodological unity in his work. As Gutting (1994: 3) puts it: 'each of Foucault's books strikes a specific tone that is muffled and distorted if we insist on harmonising it with his other books [...] his analyses are effective precisely because they are specific to the particular terrain of the discipline he is challenging, not determined by some general theory or methodology'. While Foucault outlined theoretical standpoints that seemingly build upon one another, these were always developed in relation to a rather restricted domain of investigation. Thus, in general terms, Foucault's theories and methods are 'subordinated to the tactical needs of the particular analysis at hand. They are not general engines of war that can be deployed against any target' (1994: 4); rather they are 'temporary scaffoldings, erected for a specific purpose' (1994: 16).[1]

Yet, despite his efforts to disavow any underlying world view or common theoretical perspective in his work, Foucault's various analytical standpoints have gradually congealed into clearly defined theoretical positions, especially since his death in 1984. A whole series of textbooks and primers have delineated Foucaultian perspectives on discourse, power, subjectivity, knowledge, ethics and many other topics.[2] While these delineations often overlook the contingent and provisional nature of Foucault's theoretical writings, they nevertheless bring to the fore the distinctive character of his overall approach and the implications it holds for post-structuralist analysis. Moreover, the gradual emergence of a Foucaultian 'world view' helps us to understand how Foucault's writings act to 'bridge' structuralism and post-structuralism. For instance, his early work on madness and the human sciences displayed some lingering attachments to structuralism (and Marxism) while his later work on discipline and sexuality enters more fully into the domain of post-structuralism. Thus, by briefly tracing the development of Foucault's thought, it is possible to gain a little more insight into the emergence of post-structualist theory and to assess the implications for human geography.

For this reason we concentrate here on the spatial aspects of Foucault's theory. According to Flynn (1994: 43), what is most distinctive about Foucault as a post-structuralist thinker is his sensitivity to the spatiality of history: '[h]is implicit appeal to space with its transformations and displacements [...] undermines the telic nature of traditional historical accounts, even as it refers to the dispersive character of time'. Chris Philo (2000: 218) also highlights this aspect of Foucault's work when he identifies 'spaces of dispersion' in the historical studies. These are spaces where 'things proliferate in a jumbled up manner on the same "level" as one another' thereby refuting any 'totalizing' history in which trends and processes simply unfold unproblematically through time and space. Philo argues that, for Foucault, any totalizing viewpoint 'remains alien to the details and differences of history at particular times and in particular places [...] because it inevitably smoothes over the specific confusions, contradictions and conflicts which have been the "stuff" of the lives led by "real" historical people, powerful and powerless alike' (2000: 218). Thus, instead of 'grand historical visions' we get histories of particularity and specificity, histories that attend first and foremost to local details. Importantly, in his descriptions of these 'local details', Foucault reveals quite clearly the 'materiality' of space and this aspect of his work will be the focus of our attention.

The last point to note before turning to examine Foucault's spatial histories is that he was committed to a strongly *nominalistic* approach — that is, he was interested in observing how subjects and objects come into being *in the context of* specific discursive formations. In order to follow through on this nominalism, Foucault effectively sought (in structuralist fashion) to displace the human subject from the centre of his histories. He aimed to 'account for the constitution of knowledges, discourses, domains of objects, etc., without having to make reference to a subject which is either transcendental in relation to the

field of events or runs in its empty sameness throughout the course of history'
(Foucault, 1980: 22). The historical framework that was of most interest to him
was the discursive formation – that is, the broadly distributed ways of knowing
and thinking that make up specific domains of knowledge and practice. In
Foucaultt's view, discourses help to constitute positions and perspectives that
inevitably change as discursive contexts change. He therefore frequently describes
'breakpoints' in the development of discourses and shows how these breaks
bring forth new conceptions of the world. He believes shifts in discourses produce 'new kinds of knowledge, along with new objects to know and new
modalities of power' (1980: 22). Importantly, Foucault's nominalism extends to
space: he sees spatial relations and spatial arrangements as similarly constituted
by discursive regimes of various kinds.

In what follows, we shall examine two main objects of knowledge in Foucault's
work. First, we look at the systematic knowledge of individualized subjects that
lies at the core of his studies of madness and discipline. We shall see how the
discourses of medicine and criminology 'construct' very particular subjects and
objects, and we will examine how these subjects and objects are enclosed within
new institutional spaces. Second, we shall consider how discourses of 'government' come into being that are concerned not only with the management and
sustenance of individuals but also with whole populations. As Foucault (1979: 25)
puts it: 'Governments perceived that they were not dealing simply with subjects
or even with "people", but with a "population", with its specific phenomena
and its peculiar variables: birth and death rates, life expectancy, fertility, state of
health, frequency of illness, patterns of diet and habitation'. Again, the emphasis here is on 'ways of knowing' discrete phenomena (within 'expert' disciplines
such as sociology, economics, medicine and demography) and the impact of
these ways of knowing upon 'ways of doing' in specific territorial contexts.

While we consider continuities in the means by which both individualized
subjects and whole populations are 'known', we also encounter some discontinu-
ities, most notably in the attention Foucault gives to the particularities of discrete
spatial zones as his theoretical gaze shifts from the institutional to the societal
realm. In particular, we see that the spatial mechanisms at work in Foucault's histories are more clearly evident in the early work on enclosed institutions than
they are in the later work on territorially dispersed populations. In other words,
it is argued that Foucault's work fails to move convincingly between micro- and
macro-spaces. By highlighting the problematic nature of Foucault's two 'geographies', we gain a useful insight into his general approach to 'relational space'.

Archaeology and madness

It is usual to distinguish two main periods in Foucault's intellectual career: the
first is described as the 'archaeological' period, the second as the 'genealogical'
period. These two metaphorical terms capture the substance of Foucault's concerns

throughout the two main phases of his working life. The first, archaeology, might be seen as a description of his historical method, his desire to dig beneath the surface of received 'fact' in order to divine the 'deep structures' of historical behaviour. As mentioned above, these structures are characterized as 'discursive formations'. For the most part, the study of such formations is set within those bodies of knowledge we usually refer to as 'disciplines' and the archaeological approach aims to identify how specific discourses mark the limits of what can be known at given moments in time. In his genealogical phase, Foucault is more concerned with the practical consequences of disciplines and their associated discursive frameworks. In particular, he comes to see discursive practices as part and parcel of the exercise of power. Thus, genealogical analysis explicitly links together power and discourse (or knowledge), and aims to analyse the inhibiting or constraining effects of discursive practice and the resulting impacts on social and spatial arrangements. In so doing, it takes us further into the materiality of space and the ways in which power relations shape the contours of material formations.

For the purposes of convenience we can treat the archaeological and genealogical periods as distinct. In this section we will therefore consider the main works emerging during the archaeological phase before turning to the genealogical phase in the next section. However, before putting this distinction into operation it is worth noting Eldon's caveat that archaeology and genealogy should not be viewed as mutually exclusive terms. He says:

> although genealogy is sometimes seen as a replacement for archaeology, it is better to see the two as existing together, as two halves of a complementary approach. Archaeology looks at truth as a system of ordered procedures for the production, regulation, distribution, circulation, and operation of [discourses], whilst genealogy sees truth as linked in a circular relation with systems of power which produce and sustain it, and to effects of power which it induces and it extends. (2001: 104)

On this view, the studies of discourse that emerge during Foucault's archaeological phase pave the way for the more materialistic studies of power/knowledge in the genealogical phase.

Foucault's first major work in the archaeological period concerns the history of madness (in French this was published as the *Histoire de la folie*, and in English as *Madness and Civilisation*).[3] Effectively Foucault's history documents a series of disruptions in the way insane people are treated by the rest of society. He discerns a first disruption in the mid-seventeenth century. This separates the Classical view of madness (prevalent during the seventeenth and eighteenth centuries) from the view dominating during the Middle Ages and the Renaissance. Another break can be seen at the end of the eighteenth century, and this heralds the birth of the modern view of madness. In documenting the shift from one regime of madness to another, Foucault pays particular attention to the *spaces of madness*, the places where the mad were confined and the nature of the confinement.

We can start here with the image of the 'ship of fools', a strange 'drunken boat' which emerges onto the imaginary landscape of the Renaissance. This imaginary ship, Foucault suggests, actually refers to the wanderings of the mad. During this era those people regarded as insane were driven out of towns and into the countryside where they lived an aimless and rootless existence; they moved from place to place as medieval fears defined their itinerant status. Eldon (2001: 123) summarizes it thus: 'treatment of the mad is [...] shown to be erratic – sometimes tolerant, sometimes exclusionary, sometimes hospitable. There is no regimented model, no overall plan'.

If the symbolic image of Renaissance attitudes is the ship of fools, then the symbolic image of the Classical period is the hospital. Foucault suggests that from the middle of the seventeenth century onwards the mad were no longer left to wander but were increasingly confined within dedicated 'mad spaces'. This shift takes place as part of a more general trend towards ordered urban spaces in which discrete social groups were positioned according to their function and status. Confinement was thus a new mechanism of social control within the city. Moreover, this mechanism was built around new forms of urban morality. In Gutting's view,

> the conceptual and physical exclusion of the mad reflected a moral condemnation. The moral fault, however, was not the ordinary sort, whereby a member of the human community violates one of its basic norms. Rather, madness corresponded to a radical choice that rejected humanity and the human community in toto in favour of a life of sheer (nonhuman) animality. (2001: 265)

The insane were therefore confined as a kind of moral punishment for their acquisition of inhuman characteristics and behaviours. And within these confined places, they were treated in a variety of ways: 'some had places in hospitals and almost had a medical status, whereas others were effectively in prison' (Eldon, 2001: 126).

As we enter the modern era, however, another change in perceptions of the insane takes place. Now the mad return to the human fold but are seen as offenders against social norms. As such, they require correction and treatment. Those lucky enough to be in hospital would be subject to medicalized processes of observation and classification. This medicalization of madness would lead to the condition of insanity progressively being seen not as a disease of the soul but of the body. Importantly, the re-location of madness in the space of the body allowed for the development of various medical interventions. As Gutting (2001: 266) comments: 'corresponding to this new conception of madness is the characteristic modern mode of treating the mad: not merely isolating them but making them the objects of a moral therapy that subjects them to social norms'.

Confinement and moralization combined to ensure a restructuring of space. Eldon (2001) discerns this restructuring in the architecture of the asylum, where patients were distributed, isolated and controlled. He says: 'rooms were structured so that the apparent autonomy of the patients was greater than the actual – false handles on certain doors, with some spaces forbidden; custodial features were minimised, such as the muffling of bolts, and the use of case iron frames around windows to remove the need for bars' (2001: 131). In these new asylum spaces, the mad were subjected to a moralizing judgement, a judgement that was itself closely bound into the material fabric of the spatial structure.

It should be apparent that in this work on madness Foucault weaves together morality, medicine and space to indicate how discursive formations (in this case, the discourse of madness) construct and confine human subjects (in this case, the mad). He shows a movement from a pure morality to a medicalized morality. This movement entails the construction of places of confinement where the mad are increasingly subject to the medical gaze; as this gaze is brought to bear, so madness is progressively redefined as a modern form of illness. Yet, while the doctor now occupies a central place within the asylum, 'his intervention is not made by virtue of a medical skill but by the power of morality'. Thus, the asylum ensures a new form of 'moral imprisonment' (Eldon, 2001: 133). The space of the asylum is a space of morality with the internal structure somehow reflecting the (moralistic) character of the prevailing discursive formation.

In general terms, a discursive geography of madness emerges from Foucault's historical study. He shows how 'spaces of unreason' come to be successfully demarcated from 'spaces of reason', and he illustrates how space is used in relation to the mad, 'tracing patterns of exclusion, ordering, moralisation and confinement'(Eldon, 2001: 133). The focus on exclusions and confinements thus effectively reveals the 'spaces of dispersion' identified by Philo:

> Foucault's text concerns the historical emergence in Western Europe of an impulse both social and spatial towards segregating people labelled as mad (as 'lunatic', 'insane', 'mentally ill') from the 'normal' round of work, rest, and play, often with the consequence that these people ended up living out their days in houses of confinement both non-specialist (workhouses, prisons) and specialist (asylums, mental hospitals, mental health facilities). (2000: 223)

By pointing to this outcome, Foucault's history challenges the broad thrust of Enlightenment thought, which tends to see the adoption of medicalized treatments as reflecting the emergence of a more humane attitude towards those labelled 'insane'. The shift from one episteme of madness to another is not rendered in terms of progress; rather, it is seen in terms of the introduction of a more totalizing form of confinement and moral judgement. The mad begin as wanderers and end up as prisoners. This result is fairly typical of Foucault's assessment of modern knowledge systems. According to McNay (1994: 2), Foucault generally seeks to question 'the rationality of post-Enlightenment society by focusing on the ways in which many of the enlightened practices of modernity progressively delimit rather than increase the freedom of individuals and, thereby, perpetuate social relations of inequality and oppression'.

BOX 2.1

The following issues emerge during Foucault's 'archaeological' phase:

- Space is shaped by discourse so that discursive conventions become enshrined within particular 'micro' spaces (such as the asylum).
- Actors within those spaces are 'made' by the discourses that surround them (for example, the mad are 'made' by discourses of madness).
- There are sharp breaks in the structure of discursive formations as one inevitably gives way to another.
- Breaks in discursive formations indicate that there is a residual structuralism at work in Foucault's archaeologies as the formations take on almost structural qualities – once one can 'read' the formation' one can read behaviour in micro-settings.

A concern with the 'dark underside' of progress comes explicitly to the fore in Foucault's other major archaeological work, *The Order of Things*. Here he examines how the human sciences changed during a series of shifts from the medieval through to the modern age. Again, he emphasizes the contingency of knowledge and for each period he sketches the general epistemic structure underlying the human sciences or their equivalents. Gutting explains the approach as follows:

> Foucault's characterisations of the epistemes of the Renaissance, the Classical Age, and the modern age are formulated in terms of, one, an episteme's fundamental manner of ordering the objects of thought and experience (its 'order of things'); second, the consequences of this ordering for the nature of signs (especially linguistic signs); and third, the consequences of the episteme's view of order and of signs for its conception of knowledge. (2001: 269)

As with the history of madness, this study of the human sciences delineates breakpoints, with one episteme inevitably giving way to another. For instance, Foucault argues that, within the Renaissance episteme, the underlying structure of knowledge was given by the notion of 'resemblance' in which the relation between one object and another derived from the perceived commonality of forms (for example, between signs and the things they signify). With the advent of the Classical age, this is replaced by an episteme based on the identities and differences that exist amongst objects. Thus, we witness the emergence of formal systems of signs (such as classification tables) that aim to represent the degrees of sameness and difference between things. This Classical system in turn gives way to a modern episteme in which sameness and difference come to be seen in both functional and historical terms. As Gutting

(2001: 271) puts it: 'an entity is understood and related to other things in virtue of the role it plays not in an ideal table of possibilities but in a real, historically developing environment'.

Through all this, Foucault focuses on the status of 'man'. Following his history of madness, he aims to show how human subjects are 'constructed' by disciplinary discourses – as Ian Hacking (1986) puts it, he is interested in how differing knowledge domains 'make up people'. The domains in question are psychology, sociology and literary analysis, and the focus is on the status of 'man' as a representational being. Foucault claims that in the medieval and Classical ages people were simply not capable of 'representing' the human world, because modes of representation were set within resemblances and classifications of sameness and difference. However, in the modern age, 'man' as a representational being emerges and (disciplinary) questions are asked about the type of being this 'man' might be. Again, Foucault points to the structure of disciplinary discourses in order to show how they succeed in imposing their formal structures upon diverse modes of human representation and experience. Gutting suggests that with *The Order of Things* we witness the full flowering of the archaeological method:

> Archaeology emerges as a method of analysis that reveals the intellectual structures that underlie and make possible the entire range of diverse (and often conflicting) concepts, methods, and theories characterising the thought of a given period. Concepts, methods and theories belong to the conscious life of individual subjects. By reading texts to discover not the intention of the authors but the deep structure of the language itself Foucault's archaeology goes beneath conscious life to reveal the epistemic 'unconscious' that defines and makes possible individuals' knowledge. (2001: 269)

It would seem from this comment that Foucault retains at this time a lingering connection to structuralism – that is, he continues to pay a considerable amount of attention to the underlying structures of differing discursive formations. Although in his 1970 foreword to *The Order of Things*, he berates 'half witted commentators' who persist in thinking of him as a structuralist, Eldon (2001: 101) believes Foucault is here 'protesting too much'. There are, Eldon notes, clear similarities between Foucault in his archaeological phase and the structuralists, notably in the downplaying of human agency and in the significance ascribed to formal discursive rules.[4] As Hacking (2004: 288) puts it: 'Foucault proposed his various ideas of a structure that determines discourse and action from the top down'. And yet, Eldon believes, despite these affinities, even in this period of his work, Foucault is beginning to move decisively towards post-structuralism. This is particularly evident in the attention he pays to spatiality. Eldon (2001: 102) says that Foucault's histories 'were not merely spatial in the language they used, or in the metaphors of knowledge they developed, but were also histories of spaces, and attendant to the spaces of history'. This concern for spatiality meant that Foucault was inevitably drawn to the dynamic temporality of discursive structures and their complex immersion in actual material places and spaces. We have seen some evidence of this in Foucault's history of madness but it is taken much further in the studies that comprise his genealogical phase.

Genealogy and discipline

The key text of this second period in Foucault's working life is *Discipline and Punish*, a history of penal reform and punishment, first published in 1975. The study of the prison allows Foucault to take forward themes that were explicitly addressed in *Madness and Civilisation*, notably the confinement of subjects within specific discursive regimes. However, the focus now shifts to the connections between bodies of knowledge and non-discursive practices. In particular, Foucault becomes concerned with the power relations that underpin or surround specific discourses and with the way such relations configure or construct practices of various kinds. His genealogical perspective highlights how relations of power link together discursive and material resources. Thus, the genealogical method pays particular attention to the relationship between power, knowledge, practice and space. Not surprisingly, it is during this phase that the relational character of space comes most fully into view.

As in his earlier work, Foucault contrasts the modern age of incarceration with a preceding Classical age. In the Classical period the most striking feature of the regime of punishment was its public and flamboyant character. Punishment was here a visual display of the power of the sovereign: 'pillories, gallows and scaffolds were erected in public squares or by the roadside; sometimes the corpses of the executed persons were displayed for several days near the scenes of their crimes' (Foucault, 1979: 58). Punishment thus worked as a visual medium and its power resided in its impact on the body of the miscreant, as well as in public perception of this impact. However, during the eighteenth century this Classical notion of 'punishment as spectacle' came to be questioned by penal campaigners who argued for a less physically harmful, more reformist mode of retribution. The campaigners put forward a variety of suggestions for reform, including a wider usage of exile and deportation. Yet, gradually another solution emerges: imprisonment. In Foucault's account, confinement as a mode of punishment becomes so ubiquitous that by the beginning of the nineteenth century a new regime of discipline based upon the prison has come in to being.

Just as the asylum materializes the discourse of madness, the prison materializes the discourse of crime and punishment. As Driver (1994: 283) puts it, in the modern carceral regime, individuals are to be 'trained into new habits, new patterns of conduct; their bodies subject to a dressage of disciplinary routines, their conduct monitored as closely as possible'. Activities are therefore to be strictly regulated in space and time: 'prisons are divided by cells, landings and wings, just as schools are managed by classes and hospitals by wards' (1994: 283). The internal structure of the prison comes to reflect the precepts of the modern regime of disciplinary correction. Foucault summarizes the implications thus:

> Disciplinary space tends to be divided into as many sections as there are bodies or elements to be distributed [...] its aim was to establish presences and absences, to know where and how to locate individuals, to set up useful communications, to interrupt others, to be able at each moment to supervise the conduct of each individual, to assess it, to judge it, to calculate its qualities or merits. It was a procedure, therefore, aimed at knowing mastering and using. (1979: 143)

Knowing, mastering and using, however, are further refined into a set of techniques of surveillance.

> Hierarchical, continuous and functional surveillance [...] was organised as a multiple, automatic and anonymous power [...] This enables the disciplinary power to be both absolutely indiscreet, since it is everywhere and always alert, since by its very principle it leaves no zone of shade and constantly supervises the very individuals who are entrusted with the task of supervision; and absolutely 'discreet', for it functions permanently, and largely in silence. (1979: 176–7)

The material fabric of the prison must ensure, 'hierarchical observation' – that is, 'careful monitoring by observers who are not themselves observed' (Gutting, 2001: 280). Hierarchical observation lays the groundwork for 'normalizing judgement' – that is, an assessment of prisoners that culminates in pronouncements of 'normality' or 'abnormality' (Foucault, 2004). The monitoring of bodily conduct is aimed at establishing a rigid adherence to norms on the part of prisoners so that deviant behaviour can be easily apprehended.

As Foucault explains, close monitoring requires observation by observers who are not themselves observed. He provides, as the most striking illustration of this hidden but intrusive process of observation, the example of Jeremy Bentham's Panopticon. Although it was never actually built, Foucault believes Bentham's design for the 'ideal' prison shows how 'nomalizing judgement' and 'hierarchical observation' routinely become enshrined in modern disciplinary institutions. As envisioned by Bentham, the Panopticon was a multi-storied building with a tower at the centre of a circular space (see Figure 2.1). The cells in the outer ring faced the tower with a completely open but barred frontage. The outer end of each cell was open to light from the outside so that, from the central tower, each cell was backlit in an illuminatory fashion. All activities in all cells were therefore rendered highly visible. However, the tower itself was maintained in darkness so that the prisoners could never know whether or not they were actually being watched. As a consequence, each prisoner was forced to assume *constant* surveillance even though this surveillance might be intermittent or even non-existent. In this building, then, prisoners were expected to monitor and regulate their own conduct, albeit on the assumption that they were, in turn, being monitored by an external authority. As Sharp et al. (2000: 14) summarize it: 'the upshot was the internalisation of discipline, the making of

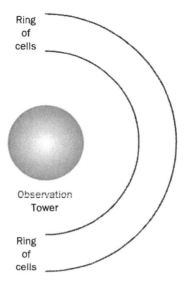

Ring
of
cells

Observation
Tower

Ring
of
cells

FIGURE 2.1 A plan of Bentham's Panopticon (Source: Hannah, 1997)

self-discipline, as inmates were enlisted into controlling themselves, and as the external eye in the inspection tower was replaced by the internal eye of conscience'. In a commentary on the Panopticon, Matthew Hannah draws out similar implications:

> Prisoners as human objects are visible as individuals: each one is distinguishable from all the others; each irregular activity is assignable to a specific person. All prisoners are potentially visible in all activities; they are completely limned by light. None can escape punishment as an automatic consequence of abnormal behaviour. The watchers in the tower have direct control of the means of punishment through a hierarchical structure of command unifying what I call three moments of control: observation, judgement and enforcement of behaviour [...] while watching is only sporadic, the threat of being watched never ceases [...] Panoptic power, then, brings together a completely visible, distinguishable and precisely punishable human object, and a unified, infallible, omniscient and anonymous authoritative subject. (1997: 347)

We clearly see in the Panopticon how power links together subjects and objects within the context of a discursive regime, one aimed at establishing specific norms of behaviour. We also see the spatial implications of this linking as complex architectural spaces bring forth the relations of power developed initially at the discursive level: 'Jeremy Bentham's Panopticon relates power and knowledge, norm and surveillance, in an interplay of architecture and social science' (Flynn, 1994: 41). By displaying these relations, the Panopticon usefully illustrates how power, discourse, practice and space come to be aligned in the regulation of prison life. It also shows how discrete spatial formations can be stabilized in relatively fixed and enduring ways.

BOX 2.2

Some core features of Foucault's genealogical approach:

- Discourse becomes deeply embedded in the materiality of given spaces to the point where it might be argued 'material arrangements' generate the 'discursive' aspects of these spaces.
- These material and discursive spaces act upon the bodies of human subjects. Thus subjectivity is constituted spatially, in some real sense it is *made* by the spatial configurations in which the subjects (that is, inmates) find themselves.
- Thus, 'external' discourses are 'internalized' to the extent that these discourses help to produce subjectivity.
- In this regard Foucault has moved much more fully into post-structuralism as the subject is now 'decentred' into the relations that surround him/her.
- Moreover, these relations combine discursive and non-discursive elements so that relational configurations can be seen as 'heterogeneous'.

Foucault focuses on the Panopticon because he believes it crystallizes key features of a new discursive regime associated with discipline and punishment. Within this regime, the penal system becomes a kind of 'factory' for producing knowledge about individual prisoners. However, this 'knowledge factory' is concerned 'not with the crimes committed but with the potential danger that lay hidden in every individual' (Barker, 1998: 56). It is therefore the suppression of potential dangers that drives the development of Panoptic monitoring and surveillance. Moreover, Foucault argues these mechanisms are increasingly adopted beyond the prison gate in a host of institutional settings – schools, factories and hospitals – where the same processes of observation and normalization are valued. Foucault claims that these new Panoptic spaces come to comprise a a 'carceral archipelago', organized in line with the ubiquitous strategies of hierarchical observation and normalizing judgement. It seems, then, that the prison has spread, heralding the emergence of what might be called a 'disciplinary society'.[5]

However, Eldon (2001) suggests there is another way of reading *Discipline and Punish*. He suggests that the Panopticon should be seen as 'the *culmination* of a variety of technologies of power rather than their *beginning*' (2001: 147; emphasis added). He believes that we should study not the Panopticon itself but 'panopticism'. In his view, 'we can best understand the birth of the prison from the general rise of what is designated panopticism, rather than the reverse' (2001: 147). The Panoptic prison is simply the pinnacle of techniques that had for long existed in the army, the school and the workshop: 'rather than institutions being a diluted form of the prison, the prison is the general trend in its most extreme form' (2001: 147). Thus, in Eldon's account, Foucault details aspects of the Panopticon not because this particular arrangement of power has gradually inserted itself into every nook and cranny of modern society but because it usefully shows how power relations in general work in 'microphysical' environments such as prisons, hospitals, schools and other institutional spaces.

Government and governmentality

The analysis that Foucault provides in *Discipline and Punish* is emblematic of his later work, especially in its focus upon strategies of 'normalization'. These strategies are of abiding interest to Foucault (see, for instance, Foucault, 2004). In fact, they are soon extended beyond micro-locales, such as prisons, to a broader study of systems of 'government' – that is, the normalization of behaviour at the *societal* scale. This turn towards 'societal government' takes Foucault beyond the juridical sphere into a host of other domains – such as education, welfare services, urban planning, economic regulation and health – anywhere that modes of 'normalizing judgement' are routinely brought to bear (2004: 134).

In order to develop his perspective on normalization 'outside' the prison Foucault adopts a very broad definition of 'government'; it applies to 'any more or less calculated and rational activity [...] that seeks to shape conduct'; that is, it applies to 'any attempt to shape with some degree of deliberation aspects of our behaviour according to particular sets of norms and for a variety of ends' (Dean, 1999: 10–11). Increasingly, then, Foucault sees disciplinary and other forms of power in terms of the 'shaping of conduct' in line with governmental strategies of 'normalization' (Foucault, 2004: 49). Yet, while this notion of the 'conduct of conduct' might be applied to almost any form of governmental activity,[6] Hindess (1996: 106) believes Foucault intends it to be applied in a narrower fashion to refer to 'less spontaneous' exercises of power over others, to 'those exercises that are more calculated and considered'. Indeed, Foucault brings government and calculation explicitly together in the notion of 'governmentality', a topic he discusses in a series of lectures in the late 1970s (eventually published in Foucault, 1991). According to Lemke (2001: 191), the development of this concept 'demonstrates Foucault's working hypothesis on the reciprocal constitution of power techniques and forms of knowledge. The semantic linking of governing ("gouverner") and modes of thought ("mentalité") indicates that it is not possible to study the technologies of power without an analysis of the political rationality underpinning them'. Thus, there are two sides to governmentality. First, the term defines a discursive field in which the exercise of power is 'rationalized'. The 'rationality' of government is defined by Colin Gordon (1991: 3) as 'a way or system of thinking about the nature of the practice of government (who can govern; what governing is; what or who is governed) capable of making some form of that activity thinkable and practicable both to its practitioners and to those upon whom it is practised'. However, as Lemke (2001: 191) notes, a political rationality is not 'pure, neutral knowledge which simply "re-presents" the governing reality; instead, it itself constitutes the intellectual processing of the reality which political technologies can then tackle'. This bring us, secondly, to technologies of government – that is, to those procedures that enable rationalities to act effectively upon diverse subjects and objects. These consist of 'mundane programmes, calculations, techniques, apparatuses, through which authorities seek to embody and give effect to governmental ambitions' (Rose and Miller, 1992: 175). Rationalities and technologies are closely aligned within specific regimes of governmentality – that is, 'thought as it becomes linked to and is embedded in technical means for the shaping and reshaping of conduct and in practices and institutions' (Dean, 1999: 18).

Foucault's work on prisons and asylums clearly indicates that the practice of government is widely dispersed throughout society. In fact, it seems that almost all forms of disciplinary expertise are being brought to bear in a governing process that extends throughout modern institutions. As Dean (1999: 10) puts it: 'there is a plurality of governing agencies and authorities, of aspects of behaviour to be governed, of norms to be invoked, of purposes sought, and of effects, outcomes and consequences'. Lemke (2001: 201) suggests that we might discern a continuum of governmentality, one that extends from political government right through to forms of individualized self-regulation. In the context of this continuum, we can see that the bulk of Foucault's work is located squarely in the middle: he analyses a range of institutional forms that sit somewhere between the state and the individualized subject.

More recent analysts of governmentality have, however, shifted the focus more firmly towards conventional notions of government. Miller and Rose (1990), for instance, argue that the notion of governmentality is particularly appropriate to understanding the conduct of political government in liberal democracies. They note that, in distinction to 'police' states (not simply current totalitarian regimes, but also the states of the pre-modern *ancien régime*), where there is an urge to specify and scrutinize all forms of behaviour, liberal democracies typically hold limits to state power. These limits have been evident since the latter years of the eighteenth century, when the term 'civil society' came to signify a realm of freedom, rights and activities *outside* the legitimate sphere of the state. Thus, the delimitation of the powers of political authorities arose in conjunction with a private, civil realm – consisting of markets, families, firms and so on – which existed d *beyond* the boundaries of the state. Simultaneously, however, government took on the role of fostering the self-organizing capacities of this civil realm: 'Political rule was given the task of shaping and nurturing that very civil society that was supposed to provide its counterweight and limit' (Rose and Miller, 1992: 179). In this endeavour, the disciplines of the human sciences had key roles to play, as Foucault was able to show in *Madness and Civilisation*, *The Order of Things* and *Discipline and Punish*.

Foucault (1991) identifies a recurring concern around the need to establish a viable boundary between state action and inaction in liberal society. In the early period of what might be termed the classical liberal state, the overwhelming assumption was that the totality of economic processes was ultimately unknowable and, as a consequence, economic sovereignty on the part of the state was impossible. Thus, any intervention by the state in this sphere had to be amply justified.[7] As in his earlier historical studies, Foucault discerns a 'breakpoint' in the middle years of the nineteenth century. Now a series of new roles for the state emerge which themselves begin to acquire something of the density and complexity formerly attributed by liberal thinkers to commercial society and the market (Gordon, 1991: 34). The economic sphere comes to be seen not just as an extant, natural state of affairs but as one that can only exist under certain political, legal and institutional conditions, and these have to be guaranteed by government. Rather than thinking of state action in terms of its necessary justification, there thus emerges 'an intimate symbiosis' (Gordon, 1991: 35) between government and civil society. The economy and, crucially, society become thought of more as a catalogue of problems *for* government than as a self-regulating sphere that can only be undermined *by* government.

As the engagement between state and society becomes more complex, so it becomes increasingly apparent that if modern governments are to manage the multiple domains of civil life they must have some understanding of these domains. The conduct of government is, then, tied to expertise, for this allows 'the calculated administration of diverse aspects of conduct through countless, often competing, local tactics of education, persuasion, inducement, management, incitement, motivation and encouragement' (Rose and Miller, 1992: 175). According to Miller and Rose (1990: 189), experts enter into a double alliance: on the one hand, they ally themselves with political authorities, translating political concerns about such issues as economic productivity, law and order, and pathology into the vocabulary of management, social science, medicine and so on; on the other hand, they form alliances with 'private' actors, translating their concerns over such issues as investment, child rearing or illness into a range of techniques for improvement. These two-way alliances result in what Dean (1999: 22) calls 'regimes of practices', which serve to define subjects and objects and codify appropriate ways of dealing with relations between them.

In short, the political governance of modern society requires a range of actors, practices and discourses to be mobilized across diverse socio-spatial domains. Political forces can only govern by influencing or co-opting domains in civil society that they do not *directly* control. Liberalism is thus marked out by the degree to which power is exercised, not so much by direct repression, but more by the invisible strategies of normalizing judgement that are brought to bear on apparently 'free' subjects (McNay, 1994). Such strategies emerge from a variety of locations including political authorities, expert institutions, media outlets and so forth. The result is a regime of governmentality in which dominant rationalities of rule somehow ensure the 'conduct of conduct' in a host of dispersed domains.

BOX 2.3

Foucault's interest in 'governmentality' gives rise to the following considerations:

- Discourses shape not just micro-spaces but much broader territories (i.e. societies) also. These governmental discourses work in much the same way as disciplinary discourses in that they configure subjectivity but now in a wider range of settings.
- Governmental discourses are made up of 'rationalities' – that is, broad justifications for governing certain spatial domains in certain ways – and 'technologies' – that is, the precise means by which rationalities can be implemented in practice..
- The combination of rationalities and technologies in the notion of 'governmentality' highlights the fact that government is a heterogeneous affair, it requires the mobilization of many resources and many differing types of actors, both 'inside' and 'outside' the state.
- The mobilization of rationalities and technologies relies upon 'expertise' of various kinds. Experts work to link governmental authorities to nominally 'free' subjects. Foucault's interest in expertise stems from his interest in disciplinary knowledges.
- Governmentalities thus work inside and outside the state and easily cross the state-non-state frontier.

The processes of confinement and discipline that were the subject of the earlier studies can now be seen as part of a broader concern for government in all its forms. As Foucault himself says of the analysis presented in *Discipline and Punish*: 'discipline was never more important or more valorised than at the moment when it became important to manage a population; the managing of a population not only concerns the collective mass of phenomena, the level of its aggregate effects, it also implies the management of a population in its depth and its details' (1991: 102). In other words, disciplinary techniques are to be seen as instruments of government. As Hindess (1996: 118) summarizes: 'the suggestion is, then, that we live in a world of disciplinary projects, many of which cut across other such projects, and all of which suffer from more or less successful attempts at resistance and evasion. The result is a disciplinary but hardly disciplined society'.

In fact, as we have already seen, liberal society is governed by a multiplicity of rationalities and techniques. As Rose and Miller (1992: 173) put it: 'political power is exercised today through a profusion of shifting alliances between diverse authorities in projects to govern a multitude of facets of economic activity, social life and individual conduct'. Discipline comprises one means of shaping conduct and regulating behaviour, but it is by no means the only one: for there are 'countless, often competing local tactics of education, persuasion, inducement, management, incitement, motivation and encouragement' (1992: 173). The analysis of discipline as presented in *Discipline and Punish* should thus be seen as but one aspect of Foucault's general analysis of government.

Subsequent studies within the governmentality field have amplified this concern for the management of territory and therefore space. For instance, Murdoch and Ward (1997) investigate how statistical representations of territory allowed the British state to bring conceptions of a national rural space into being in the eighteenth and nineteenth centuries. This national space was super-imposed upon the many local rural formations that could previously be found scattered throughout the British countryside. As powerful modes of national representation emerged so a national territory was consolidated in government policy. By the middle years of the twentieth century this national territory was represented as a 'national farm', a spatial zone that would be administered by state agencies in line with the governmental priority of increased food production. The effect was a radically reconfigured spatial assemblage in the British countryside (for example, larger farms, fewer farmers, more machinery and a changed natural environment).

Perhaps because they assess such spatially extensive entities as the countryside, the rural and the agricultural, Murdoch and Ward can stress the statistical emergence of territory and its gradual solidification within governmental modes of representation. However, in using the same perspective to analyse the nineteenth-century city, Osborne and Rose (1999: 740) emphasize a rather different set of governmental concerns. They believe the city at this time must be seen as 'a plane of indetermination – a dense, opaque, unknown, perhaps ultimately unknowable space: a domain where the criteria and techniques of good government were no longer self-evident'. Where Murdoch and Ward's account of rural governmentality stresses the effective and far-reaching nature of governmental interventions, Osborne and Rose see urban government as 'having ambitions that were entirely negative', linked to fears of the mob, problems of overcrowding and the degenerating effects of urban squalor. In assessing governmental responses to these problems, Osborne and Rose (1999: 758) conclude that there is something 'ungovernable' about the city, as efforts to convert the sociability of the city to the ends of government appear to simultaneously require the preservation of the 'spontaneous underdetermined character of the city itself'. Governmental distinctions between country and city therefore seemingly rehearse the age-old problem of liberalism – that is, where to draw the line between state and society (for a fuller discussion of urban–rural distinctions see Chapter 7).

It would appear, then, that Foucault's conceptualization of government can help us to understand the relationship between space (for instance, in the form of institution or territory) and discourse (for instance, in the form of differing mentalities and techniques of rule). However, we should note at this juncture that the shift from the micro-level of the asylum and the prison to the macro-level of societies and states is not achieved quite as seamlessly as the account given above might suggest. For instance, John Allen (2003: 75) has expressed some disquiet about the move from one scale of geographical analysis to another in Foucault's genealogical studies. He believes the 'diffuse topography' evident in the governmentality literature 'sits rather awkwardly next to the meticulous and rather dense configurations of the prison or the clinic'. Allen argues that once Foucaultian analysis moves beyond particular sites and specific institutions, it tends to become 'impressionistic' and 'metaphorical' – that is, it loses sight of the precise spatial arrangements (detailed in *Madness and Civilisation* and *Discipline and Punish*) that obviously work to regulate behaviour. He says:

> In contrast to the detailed survey of techniques in Foucault's earlier institutional analyses – documenting the distribution of individuals in penal spaces, for example, on the basis of a series of grid-like expectations about how prisoners should conduct themselves – we have scant detail of the spatial assemblages involved in the management of dispersed populations. (2003: 82)

In other words, as Foucault's gaze shifts from enclosed micro-spaces to more diffuse macro-spaces, the specifically spatial aspects of his approach fade into the background.[8]

In Allen's view, Foucault's geography can be seen most clearly when the precise 'diagrams of power' encoded in institutions like the prison and the asylum are shown to interact with broader discursive formations or modes of classification: 'the layout, disposition and orientation of the various clinical or prison buildings [...] are all deemed to have played a part in inducing particular forms of conduct, although not in isolation from the classificatory techniques and normative strategies designed to engage the minds of particular subjects' (Allen, 2003: 71). Allen goes on to say:

> [R]egular forms of conduct are indeed induced, but not because they are 'read off' by subjects from a particular series of techniques or a particularly stark spatial arrangement. Rather it is the interplay of forces within a particular setting which makes it possible to extrapolate diagrams from the power relations inscribed within particular institutional spaces: subjects are progressively constituted, symbolically and practically, through specific points of purchase; mobilised and positioned through particular embedded practices; and channelled and directed by a series of grid-like expectations about how, when and where to conduct themselves and others. In simple terms, different kinds of diagrams make different kinds of government and control possible, even though things rarely turn out quite as planned. (2003: 73)

In Foucault's institutional studies the diagrams are easy to see and it is clear that spatial and discursive arrangements become intimately intertwined as behaviours are regulated and as practices are moulded by governing agencies (even though, as Allen emphasizes, 'things rarely turn out quite as planned', a point that we shall consider at some length in subsequent chapters). In the institutional micro-locales, the linkage between spatial relation and spatial formation is easy to see. Yet, with the shift of emphasis to widely dispersed processes of government, we lose the rich spatial vocabulary of the institutional diagrams: 'once outside the walls of the institution, so to speak, it was as if a concern for the detailed spacing and timing of activities, and how they induced and channelled particular patterns of behaviour, no longer had any real purchase on the more expansive matters at hand' (Allen, 2003: 90). This results, Allen argues, in a 'geographically skewed topology' in which 'the transformation of power relations across space is of less fascination or interest than those transformed in space' (2003: 89).

Power and space

Allen's reservations about the shift in the spatial focus of Foucault's genealogical studies are expressed during a general consideration of Foucault's work on power. In Allen's view, Foucault is the pre-eminent exponent (along with Gilles Deleuze, see Chapter 3) of the notion that power is an 'immanent affair' – that is, power is a normalizing force, one that works *through* (rather than *upon*) the discourses, techniques, practices and arrangements which frame and compose everyday life.[9] This perspective on power has proved highly influential and is usually seen as a central aspect of Foucault's contribution to post-structuralist thought (McNay, 1994; Hindess, 1996). In this section, I will briefly outline Foucault's account of power before returning to Allen's reservations about its application in spatial analysis.

Power became increasingly central to Foucault's work in his later years, especially in the studies of discipline and government. In Flynn's (1994: 34) view, 'power relations underwrite all Foucault's genealogies'. In the genealogical phase, Foucault ties together knowledge, discourse, space and power, with power relations acting to somehow bind all these aspects together. For instance, in his studies of the asylum and the prison, Foucault shows how power works *through* discursive regimes, spatial arrangements and social practices. He also shows how the patients and the prisoners are made the subjects of power; that is, he shows how they become *subjected to* power relations of various kinds.

In *Discipline and Punish*, the mode of subjection is discipline: 'Discipline makes individuals; it is the specific technique of power that regards individuals both as objects and as instruments of its exercise' (Foucault, 1979: 170). Power is here immanently invoked within the range of detailed techniques – hierarchical observation, normalizing judgement and examination – that comprise the disciplinary regime. Foucault refers to these techniques as the 'micro-physics of power' (1979: 26), and he discerns micro-physical power relations in the instruments, techniques and procedures that are brought to bear within the confines of the prison. In other words, the prison can be seen as an assemblage in which relations of power are organized in a hierarchical fashion.

BOX 2.4

Some general features that characterize Foucault's perspective on power:

- Power and knowledge are closely combined.
- Power relations are interwoven with social practices and material arrangements.
- Knowledge and practice construct a world that is both knowable and governable.
- Power/knowledge relations produce subjects whose behaviour is regulated and modified in line with given rationalities.
- Power circulates through specific assemblages of materials and practices.
- Power produces a series of local effects within these assemblages.

In this broad characterization of power we can also see some general features of Foucault's spatial sensibility:

- Power works through knowledge domains that specify how particular sites should be organized.
- Modes of spatial organization simultaneously constitute power/knowledge relations.
- There is no clear distinction between power, knowledge, practice and space – all these aspects are interwoven with one another.
- This interweaving shows space to be relational in nature.

These general observations are enough to show that Foucault sees power almost everywhere. And he sees power almost everywhere because he believes it comes from almost everywhere – discourse, knowledge, practice, spaces of dispersion and so forth. And yet, as Allen (2003) observes, despite the obviously diffuse nature of power relations in Foucaultian theory, it is only within enclosed institutional sites such as asylums and prisons that Foucault seems able to successfully reveal the spatial mechanisms at work. Moreover, it is also clear that within such sites power is both *dispersed* (for instance, in materials, techniques and practices) and *concentrated* (for instance, in processes of observation and surveillance). In fact, the degree of repression and prohibition evident in Bentham's Panopticon suggests that, for the most part, *Discipline and Punish* refers to a state of *domination* in which the prisoners are reduced to the status of 'docile bodies'. As McNay notes:

> Foucault's analysis of the disciplinary techniques within the penal system is skewed towards the official representatives of the institutions – the governors, the architects, etc. – and not towards the voices and bodies of those being controlled. Failure to take

account of any 'other' knowledges – such as prison subculture or customs inherited from the past – which those in control may have encountered and come into conflict with means that Foucault significantly overestimates the effectiveness of disciplinary forms of control. (1994: 101)

In McNay's view, this partial perspective means that Foucault 'slips too easily from describing power as a tendency within modern forms of social control to positing disciplinary power as a fully installed monolithic force which saturates all social relations' (1994: 104). We ultimately gain, then, a rather traditional view of power as the ability of a regime to exercise control *over* its subjects. In *Discipline and Punish*, to be a subject of power is quite clearly to be *subjected* to power of a prohibitive and repressive kind.

However, upon completing *Discipline and Punish*, it is clear that Foucault came to realize he had described power rather too negatively in that volume. For instance, in lectures he gave in 1975 (eventually published in Foucault, 2004) he explicitly argued a need to escape from 'outdated historical models' that see power as always 'prohibiting, preventing and isolating' (2004: 51).[10] Instead he asserted that 'what the eighteenth century established through the "discipline of normalisation", or the system of "discipline-normalisation" [is] a power that is not in fact repressive but productive, repression figuring only as a lateral or secondary effect with regard to its central creative and productive mechanisms' (2004: 52). Thus, discipline repression is superimposed upon 'positive techniques of intervention and transformation' (2004: 50). These 'positive' techniques are seen simply as 'government', meaning the 'conduct of conduct'. We can thus discern two main types of power relation: one (in the prison) that is dominant and coercive; another (in processes of liberal government) that is productive and affirmative. As Judith Butler explains,

> We are used to thinking of power as what presses on the subject from the outside, as what subordinates, sets underneath, and relegates to a lower order. But if, following Foucault, we understand power as forming the subject as well, as providing the very condition of its existence and the trajectory of its desire, then power is not simply what we oppose but also, in a strong sense, what we depend on for our existence and what we harbour and preserve in the beings that we are. The customary model for understanding this process goes as follows: power imposes itself on us, and weakened by its force, we come to internalise or accept its terms. What such an account fails to note, however, is that the 'we' who accept such terms are fundamentally dependent on those terms for 'our' existence. Are there not discursive conditions for the articulation of any 'we'? Subjection consists precisely in this fundamental dependency on a discourse we never chose but that, paradoxically, initiates and sustains our very agency. (1997: 2)

This second, positive, mode of power comes to the fore in Foucault's last publications, notably the three volumes that comprise his *History of Sexuality*. Here, power is still thought of as 'the total structure of actions', but this structure bears upon the actions of subjects that are free to choose among alternative courses of action (Foucault, 1982: 220). As Hindess (1996: 100–1) puts it:

'power is exercised over those who are in a position to choose and it aims to influence what their choices will be [...] where there is no possibility of resistance there can be no relation of power'. Foucault (1982: 213) refers to this interaction between power and resistance as an 'agonism' – that is, 'a relationship which is at the same time reciprocal incitation and struggle'. Thus, as Paul Patton points out, the human material that systems of power work upon is not docile but active:

> it is composed of forces or endowed with certain capacities. As such it must be understood in terms of power, where this term is understood in its primary sense of capacity to do certain things [...] whatever else it may be, the human subject is a being endowed with certain capacities. It is a subject of power, but this power is only realised in and through the diversity of bodily capacities and forms of subjectivity. (1998: 65)

We here arrive at a conception of a subject that is not just subjected to (negative) power relations but also actively constructs (positive) power relations. Moreover, this subject is also *embodied*; thus, power relations work upon and through bodies while resistance to power also takes an embodied form. This naturally leads on to a concern for the spaces of embodiment including the prison, the asylum and so forth.

In seeking to understand more fully the productive subject of power, Foucault begins to look more closely at 'sources of selfhood'. In particular, he discusses 'technologies of the self' – that is, the ways in which individuals 'effect by their own means or with the help of others a certain number of operations on their own bodies and souls, thoughts, conduct and ways of being, so as to transform themselves' (1988: 18). Foucault also begins to investigate an 'ethics of the self'. This involves not only a relationship to oneself as an ethical or moral agent but also recognition of oneself as a subject of power relations of various kinds. As Arnold Davidson (1994: 119) notes: 'in his last writings Foucault expressed concern that the ancient principle "know thyself" had obscured, at least for us moderns, the similarly ancient requirement that we occupy ourselves with ourselves, that we care for ourselves'. In these late works (for instance, *The History of Sexuality*, volumes 2 and 3) Foucault sees the self as something to be worked on *by the self*. In other words, the subject is no longer simply subjected to constraining power relations but can operate within productive relations to fashion new ways of being.[11]

We therefore arrive at the position where power is always exercised between subjects that have (to varying degrees) their own powers. As Pottage (1998: 23) puts it: 'power presupposes freedom in the sense that the relation itself is sculpted by a constant movement of reciprocal anticipations and interventions such that each actor is dependent on the autonomy of the other'. In the context of this 'constant movement' power relations are always potentially resistible and reversible; stability (especially in the form of domination) is not easy to achieve. Thus, in Patton's (1998: 68) view, 'it is only in exceptional circumstances that A can be sure of achieving the desired effect on B. Only when the possibility of effective resistance has been removed does the power relation between two subjects become unilateral and one-sided'. *Discipline and Punish* therefore describes not the routine imposition of power relations but a fairly extreme version of power as repression. In more normal circumstances, power relations sit somewhere between domination and freeplay: they comprise mixtures of the negative and the positive. This is perhaps most noticeable in the case of liberal government as it 'hovers between forbidding subjects/objects on the one hand and constituting objects/subjects on the other' (Barker, 1998: 66). The balance between direction and constitution entails that power relations be both flexible and robust, amorphous and consolidated. It ensures also that the existence of power relations through space and time 'depends on a multiplicity of points of resistance: these points play the role of adversary, target, support or handle in power relations. These points of resistance are present everywhere in the power network' (Foucault, 1981: 95).

BOX 2.5

On power, Foucault concludes that:

- Power is dispersed across many heterogeneous domains with many of these domains retaining their own specific powers (which can be realized in strategies of either resistance or accommodation to hegemonic forces).
- The construction and consolidation of power relations takes a considerable amount of work and the work increases as resistance increases.
- At times the consolidation and imposition of power relations can result in domination (as in the panoptic prison).
- But perhaps more routinely power leads to the production of new forms of subjectivity (as in processes of liberal government).
- Thus, Foucault's work helps us to see the diversity of power relations and their effects.

In actual fact, Foucault shows power relations to be *so* diverse that our attention is increasingly drawn not to power *per se* but to the 'materials' that make power whatever it 'is'.[12] And given that power is constituted through bodies, practices, spaces and so forth, it cannot be seen as something imposed from above or from the outside; rather, as John Allen (2003: 9) remarks, it is 'coextensive with its field of operation'; this field of operation arranges materials, demarcates spaces and produces (various forms of) subjectivity. As Deleuze says during a commentary on Foucault's work:

these power-relations, which are simultaneously local, unstable and diffuse, do not emanate from a central point or unique locus of sovereignty, but at each moment move from one point to another in a field of forces, marking inflections, resistances, twists and turns [...] There is a multiplicity of local and partial integrations, each one entertaining an affinity with certain relations or particular points. (1988: 73–4))

Thus, a relational view of power brings us again to a relational view of space. Space is here composed by the variable construction and consolidation of power relations. Discrete spaces emerge out of complex assemblages of discourses, practices and materials, all somehow bound together by relations of power. Moreover, power is not only materialized in space, it is also 'localized': it works relationally through situated and specific knowledges, practices and materials, all arranged at precise points and bound together by heterogeneous actions of alignment. As Foucault (1986: 252) says: 'space is fundamental in any exercise of power'.

Yet, if we return once again to the distinction between 'micro-physical' and 'macro-physical' forms of power (loosely associated in Foucault's work with, on the one hand, the closed institution and, on the other, liberal government) then we see that the 'localized' (and therefore spatialized) character of power relations can all too easily get lost in the move from the smaller to the larger scale. This point is clearly expressed by Allen when he suggests that the 'expansive and diffuse topography' of governmental modes of power compares unfavourably with 'the rather dense configurations of the prison or the clinic' (2003: 75). Where, in the institutional setting, we have clear descriptions of the spatial arrangements that reflect precise configurations of power, at the level of government we get 'scant detail of the spatial assemblages involved in the management of dispersed populations' (2003: 82). Allen is concerned that the mechanisms which allow power relations to be assembled *inside* institutional spaces are much easier to see than the mechanisms that allow power relations to be assembled *across* non-institutional spaces. In short, Foucault describes 'spaces of domination' much more convincingly than 'spaces of production'. Allen believes this problem is exacerbated by the fact that we cannot simply 'aggregate up' the institutional mechanisms and techniques described in books such as *Discipline and Punish* to the level of a society or a state. As he says: 'Bridging the gap between here and there to bring a diffuse population within reach is singularly unlikely on the basis of a scaled-up version of confined arrangements' (2003: 84). Allen therefore concludes that 'the challenge for those who hold that power has an immanent presence is to grasp how, in the context of a diffuse population composed of a multitude of wills, the subject and power remain mutually constitutive of each other in space and time' (2003: 85). In order to fully meet this challenge, he explains, Foucaultian scholars must turn their attention to the spatially mediated relationships that compose modern systems of governmental power in order to show how relations are stabilized across heterogeneous spaces. In other words, we need to attend to the precise mechanisms that allow spatially dispersed and seemingly autonomous and independent subjects to be aligned with particular strategies of discipline and normalization. We need, in short, to attend to the constitutive as well as the coercive powers of governmental space.

Conclusion

The reflection on Foucault's work provided above shows that the notion of 'relational space' emerges strongly from within his studies of discourse, knowledge and power. However, Foucault also focuses our attention on the interrelationship between spatial relation and spatial formation: he shows that particular discourses, networks of power, sets of material resources can all be stabilized in discrete spatial zones (the hospital, the prison and so forth). The spatial fabric of given institutions is, then, a key means of 'materializing' discursive relations. The space of the prison and the space of the asylum serve to 'perform' the relations of power specified at the discursive level.

Space and power mutually constitute one another in Foucault's work. Yet, the nature of this constitution gets harder to discern once we move out of the enclosed institutions into the dispersed populations of nation-states and other large-scale political units. Foucault clearly feels the spatial mechanisms at work are much the same: governmental processes of discipline and normalization act to configure modes of subjectivity and serve to regulate patterns of behaviour; they work in amorphous and dispersed ways and are multiple in form (the system of domination discerned in the Panoptic prison is merely at one end of a continuum of power relations). In the eyes of critics, however, the shift away from micro-scale power relations entails a loss of spatial focus: the precise means whereby dispersed and diverse relations of power act upon dispersed and diverse forms of subjectivity become hard to discern. Instead of the seamless integration of power, knowledge and space we get the assertion of governmental discourses that seemingly work both everywhere and nowhere.

We can therefore conclude that Foucault's work takes us some considerable distance in our exploration of relational space but not quite far enough. His analyses of the asylum and the prison are exemplary in helping us to understand how institutional spaces come to reflect particular power/knowledge configurations. But his extension of this analysis beyond enclosed institutions raises certain questions about the relationship between power/knowledge systems and the wider dispersal of these systems across extensive territories. In order to investigate the dispersal of power within diverse and loosely co-ordinated spatial arrangements, we turn in the next chapter to examine in a little more detail the precise spatial mechanisms that must be employed if heterogeneous spaces are to be aligned by 'rationalities of rule'. In so doing, we investigate more fully how power becomes 'materialized' in 'things' and how, in the exercise of power relations, the alignment of things becomes as significant as the alignment of people. Importantly, it is in this mixture of people and things that we begin to appreciate the full extent of relational space.

SUMMARY

In this chapter we have investigated key aspects of Foucault's work and have attempted to draw out the implications for spatial analysis. It was shown that Foucault worked for most of his career with an implicit notion of relational space, but this only came fully to the fore in his 'genealogical' phase with books such as *Discipline and Punish* and the essays on governmentality. In his analysis of power Foucault clearly shows how the social and the spatial are bound inextricably together – the one is 'immanent' in the other. This sets the scene for the fuller investigation of relationalism undertaken by other post-structuralist authors.

FURTHER READING

There are many books on Foucault, but few deal explicitly with his work on spatiality. For a general introduction Lois McNay's *Foucault: A Critical Introduction* is one of the best. For an interesting reflection on Foucault's spatial thinking, see Stuart Eldon's (2001) book, *Mapping the Present: Heidegger, Foucault and the Project of a Spatial History*. For an overview of work on governmentality, Mitchell Dean's (1999) book, *Governmentality*, is a useful starting point. For an excellent analysis of Foucault's work on power, with particular relevance to geographers see John Allen's (2003) *Lost Geographies of Power*.

Notes

1. We should also acknowledge that Foucault's interests, while mainly historical, always relate to some issue of pressing contemporary concern. He suggests his writings might be termed 'histories of the present' (1979: 30–1), in that they attempt to reveal how current circumstances *could have been different*. Thus, Foucault's histories aim to show that the processes leading to our present practices and institutions were by no means preordained or inevitable. This focus on 'histories of the present' again highlights the specific and unique nature of Foucault's historical writings.

2. For recent examples see McNay (1994), Barker, (1998), Dean (1999), Danaher et al. (2000), Miles (2003).

3. It should be noted that *Madness and Civilisation* contains only around one half of the original French text. For this reason, various commentators have argued that only the original will suffice but as this has yet to be translated into English I will use the English version as well as commentaries on the French version.

4. As Darier (1999: 13) puts it: 'Foucault may have tried to turn his back to structuralism, but structuralism remains stuck to his back'.

5. Sometimes this disciplinary society seems confined to certain key institutional spaces and at other times it seems all-encompassing. In line with this latter view, Foucault (1979: 227) makes comments in the following vein: 'the ideal point of penality today would be an indefinite discipline: an interrogation without end, an

investigation that would be extended without limit to a meticulous and ever more analytical observation, a judgement that would at the same time be the constitution of a file that was never closed, the calculated leniency of a penalty that would be interlaced with the ruthless curiosity of an examination'. The tendency then is to greater and greater surveillance with few clues as to how this tendency might be resisted.

6. Foucault adopts this broad and all-encompassing notion of government because he is referring back to eighteenth-century meanings of the term associated with philosophy, medicine, guidance for the family and so on (see Lemke, 2001).

7. However, before concluding that the demarcation line between state and civil society was firmly drawn at this time it should be noted that 'laissez faire is a way of acting, as well as a way of not acting' (Gordon, 1991: 17): it is both a limitation on political sovereignty and a positive justification for market or civil freedom.

8. Allen here echoes Massey's (1992: 80) complaint that Foucault proposes 'a notion of space as instantaneous connections between things at one moment'.

9. Andrew Sayer suggests that immanence should be thought of as 'emergence':

> Where there are two or more objects in an internal relation, that is one in which the nature of each of the *relata* depends on the other(s) through their relationship itself, instead of merely being contingently or externally related, it is possible for them to develop "emergent powers". These are causal powers dependent on but irreducible to those of their constituent elements, just as water has emergent powers for those of its constituents, hydrogen and oxygen. (2004: 266)

10. Foucault has in mind here the continued existence of models based upon slave society, caste society, feudalism and the administrative monarchy. He says, the continued use of such models comprises 'a failure to grasp what is specific and new in what took place during the eighteenth century and the Classical Age' (Foucault, 2004: 51).

11. It is worth noting that this turn to 'selfhood' in Foucault's last years has been treated sceptically by some critics. For instance, Christopher Norris (1994: 160) argues that even in his last writings Foucault continues to see subjectivity as 'constructed through and through by the various discourses, conventions or regulative codes that alone provide a means of "esthetic" self-fashioning in the absence of any normative standard'. He thus argues that

> what emerges is not as much a radical rethinking of [the] issues as a shift in rhetorical strategy, one that allows [Foucault] to place more emphasis on the active, self-shaping, volitional aspects of human conduct and thought, but that signally fails to explain how such impulses could ever arise, given the self's inescapable subjection to a range of pre-existing disciplinary codes and imperatives that between them determine the very shape and limits of its "freedom". (1994: 161)

Likewise, Butler (1997) remarks that Foucault failed to elaborate on the specific mechanisms that help to form specific subjects. These criticisms might be taken to indicate that Foucault was never able to fully extricate himself from structuralism.

12. In a similar vein, Deleuze (1988: 25) says in his commentary on Foucault's legacy: 'power is not homogeneous but can be defined only through the particular points through which it passes'.

3

Spaces of heterogeneous association

Look upon it this way: the search for pattern is an attempt to tell stories about ordering that connect together local outcomes. (Law, 1994)

Introduction

As we have seen in Chapter 2, Foucault portrays space as intrinsic to discursive regimes. Within such regimes, power, knowledge and space mutually compose one another. As power relations come into being, discourses, knowledges and spaces gain shape – they co-evolve in complex ways, coiling around one another until some kind of stability emerges. Thus, within these heterogeneous assemblages any separation of the discursive and the spatial becomes almost impossible to conceive: knowledge is materialized in practice, practice is materialized in the body, and the body is immersed in modes of spatial organization that in turn 'perform' systems of knowledge. Foucault conjures up this circular assemblage of power most clearly in his description of the (Panoptic) prison. Here, systems of knowledge bring together hierarchical observation and normalizing judgement within a regime of disciplinary power. This power extends beyond the realm of knowledge into architectural arrangements, which are designed to allow observation, judgement, regulation and normalization to occur on a regular basis. In short, the prison emerges as a stable and coherent entity from the confluence of discourse, practice and spatial organization.

Foucault's analysis of the prison illustrates the crucial role that space plays in the construction of power relations and the crucial role that power relations play in the construction of space. He shows how the composition of given micro-spaces follows from the discursive and material constitution of given assemblages of power. Yet, as we also saw in Chapter 2, the circulation of power relations *beyond* enclosed institutions is nowhere fully explained by Foucault. While he uses the term 'government' to describe broader alignments of power, knowledge and practice, he fails to investigate the spatial mechanisms at work in such alignments. Within the prison, the integration of social and spatial elements is clear to see: but outside the prison, the spatial dimension drops from view; we simply find power relations (that is, Panopticism) circulating in a kind of spatial vacuum.

If we are to build on Foucault's insights we must move outside institutions such as prisons and asylums to spaces that are co-ordinated on a more extensive basis. In other words, we need to go beyond the enclosed institutions in order to consider how power circulates *between* clearly demarcated sites. In this chapter, we undertake this task. The aim is to show how spatial scales come to be aligned with one another by relations that somehow move 'upwards' from the local level and 'downwards' from larger spatial scales. By investigating the precise ways that such alignments emerge, we hope to illustrate how social relations of various kinds are extended across space and how these relations give rise to differing spatial scales.

In taking the analysis forward in this way, we make an important move beyond Foucault's concern only for the *human* sciences – that is, we show how his general approach can be brought to bear on the *natural* sciences. In so doing, we consider 'post-positivist' accounts of scientific activity that discern a close association between power and scientific knowledge (in much the same way that Foucault himself discerns a close association between power and knowledge in the human sciences). These 'post-positivist' approaches tend to reject the view that science gains its power from its *accuracy* – that is, from its direct observation of the way the world 'really is'; rather, they see the power of science lying in its ability to control and manipulate elements, both human and natural, in ways that allow scientific facts to be built and then disseminated beyond the centres of scientific practice. In this view, 'power is no longer external to [scientific] knowledge or opposed to it; power itself becomes a mark of knowledge' (Rouse, 1987: 19).

The main focus of the chapter is 'actor-network theory', an influential perspective on scientific knowledge that has been developed over the last twenty years or so by a trio of sociologists – Bruno Latour, Michel Callon and John Law. The approach takes Foucault's observations on power/knowledge as a starting point but builds upon these in order to account for the extensive power of science and technology in contemporary society. In parallel with Foucault's focus on paradigmatic sites, actor-network theory sees the laboratory (as opposed to the prison) as the crucial citadel of power in the modern world. In Foucaultian fashion, 'the laboratory, like the clinic, the asylum, the school, the factory, and the prison, serves as one of the 'blocks' within which [...] a 'micro-physics of power' is developed and from which that power extends to invest the surrounding world' (Rouse, 1987: 107). Yet, while it pays a great deal of attention to the internal organization of the laboratory, actor-network theory's main interest is in the relationship between the laboratory and its external environment. In other words, the actor-network approach focuses on the means whereby laboratories draw entities in from the outside, subject them to various processes of transformation, and then export them to the rest of the world in the form of scientific facts and artefacts. It is this concern for the relationships between laboratories and other, external, micro-locales that allows actor-network theory to elucidate the various mechanisms that tie locations together across space.

We should, however, note that while actor-network theory clearly originates, at least in part, from Foucault's work on the human sciences, over time it has gradually moved away from a concern both for the laboratory and for power. In many ways the theory makes the most of the Foucaultian insight that it is not power *per se* that is important but the various materials, practices, discourses in which power relations are both *embedded* and *transported*. The theory therefore increasingly focuses on the complex alignments of heterogeneous entities that allow powerful scientific networks to emerge into the world. These networks are thought to link laboratories to chains of actors in a variety of other spatial locations. Thus, actor-network theory spends a great deal of time examining how actors are incorporated into chains and networks. In so doing, it also indicates how discrete spaces come to be relationally linked together. It thus shows how (networked) relations constitute and compose differing spatial locations. In particular, it investigates how processes of spatial demarcation (that is, 'localization') take place *within* network formations. In so doing, it introduces another aspect of relationality: the way spatial distinctions are carved out of broader social contexts, in this case networks.

In what follows we firstly examine the origins of actor-network theory and show how it emerged from the social scientific studies of laboratories undertaken during the late 1970s and early 1980s. We then turn to examine how the theory conceptualizes the relations between actors and spaces – that is, how it comes to adopt the notion of the 'actor-network'. Having assessed actor-network theory's distinctive approach to relationality, we then move on to consider some of the broader implications of the theory, notably its focus on 'hybrid' networks in which people and things get relentlessly 'mixed up'. Finally, we reflect on the status of the theory and tease out some of the main implications for understandings of relational space. As we shall see, actor-network theory poses some significant challenges to taken-for-granted notions of geography and space. These challenges will be assessed over the course of the next two chapters.

The emergence of actor-network theory

In the 1970s, following Kuhn's (1962) penetrating critique of positivist conceptions of scientific knowledge, a group of sociologists ventured into the citadels of scientific activity – laboratories – in order to study scientists at work. Their aim 'was to create a legitimate space for sociology where none had previously been permitted, in the interpretation or explanation of scientific knowledge' (Shapin, 1995: 297). The resulting ethnographic studies dealt a further blow to the still influential assumption that there is some simple correspondence between scientific knowledge and nature.[1] Within the ethnographies, scientists are shown to be using a variety of means to bring nature 'into being' in the laboratory just as Foucault had shown the human sciences bringing particular conceptions of 'man' into being within prisons and asylums (Hacking, 1986). The means include inscription devices, which serve to transform natural materials into literary techniques of persuasion (Latour and Woolgar, 1979), and political strategies, which permit the building of coalitions in favour of some scientific research programmes over others (Knorr-Cetina, 1981).

The laboratory studies seemed to demonstrate that scientists, far from simply observing nature, are busy actively constructing natural entities using all the social, economic and technological tools at their disposal. As Karin Knorr-Cetina (1981: 152) puts it, 'the study of laboratories has brought to the fore the full spectrum of activities involved in the production of knowledge. It has shown that scientific objects are not only "technically" manufactured in laboratories but also inextricably symbolically and politically constructed'. Thus, the laboratory studies emphasized that scientific outcomes (facts and artefacts) result from complex *social* processes. And the discovery of these social processes further undermined the rather simplistic understanding of scientific endeavour proposed in positivist accounts (see Zammito, 2004, for a discussion).

In the wake of the laboratory studies, scientific knowledge became a legitimate topic of sociological investigation. Thus, the content of science could be examined from a sociological perspective, with notions such as power, interest, norm, gender and class all being used to account for scientific behaviour (see Barnes et al., 1996, for an overview of this work). As a result, laboratories came to be seen as little different from other social settings and scientists came to appear much like other social actors. However, this finding raised a problem that Bruno Latour, himself a pioneer of the laboratory study (see Latour and Woolgar, 1979), took as a starting point for developing a rather distinctive mode of analysis. He began by asking: 'if nothing scientific is happening in laboratories, why are there laboratories to begin with and why, strangely enough, is the society surrounding them paying for these places where nothing special is happening?' (Latour, 1983: 141–2). By posing this question, Latour was expressing a concern that the social studies of science, in questioning many of the presumed special attributes of scientific knowledge generation, had also begun to undermine the sociologist's ability to account for the *power of* modern science. As he began to address this concern, Latour started to take the laboratory study in a new direction, a direction that led ultimately to 'actor-network theory'.

Latour (1983) approaches the task of accounting for science's power in the world through the use of a case study, which will be briefly summarized here. The case begins in 1881 with Louis Pasteur at work in his laboratory in the Ecole Normale Superieure in Paris. Pasteur at this time had managed to arouse the interest of French society in his experiments – that is, he had become an influential, perhaps even a great, scientist. Latour asks, how was this so? How could Pasteur gain so much support from non-scientists? In answering these questions, Latour claims that Pasteur used a tried-and-tested approach: in short, 'he transfers himself and his laboratory into the midst of a world untouched by laboratory science' (1983: 144). In this case, it is the world of anthrax, a problem causing a great deal of distress in France at the time. The first move that Pasteur makes is to establish a link between laboratory and field. He does this by constructing a makeshift lab on a farm site in order to study the anthrax bacillus. Here he 'extracts', 'treats', 'filters' and 'dissolves' materials in order to render the bacillus visible (see Latour, 1999). Having completed the on-farm study, he then makes a second move and transfers the lab back to the Ecole Normale Superieure, taking the bacillus with him. According to Latour (1983: 146), Pasteur 'a master of one technique of farming that no farmer knows, microbe farming. This is enough to do what no farmer could ever have done: grow the bacillus in isolation and in such a large quantity that, although invisible, it becomes visible'. Once this move is accomplished, Pasteur suddenly gains the ability to talk with great authority about the anthrax bacillus, especially after he shows that it causes anthrax, a problem of considerable significance in French agriculture.

At this stage, however, the 'cause' of anthrax is still locked up inside Pasteur's laboratory and it has no real bearing upon either the disease or French society as a whole. The connections between the laboratory and all those potentially interested in Pasteur's work are weak and might easily be broken apart. If this situation prevails, Pasteur's power to interest society in general will be severely limited. Thus, it is necessary for Pasteur to make another move – from the laboratory back to the field. Having manipulated the bacillus in the lab, he manages to refine a vaccine which can then be submitted to a field trial. However, Pasteur is here confronted with the problem of ensuring effective vaccination procedures. How can such procedures be put in place? According to Latour (1983: 151–2), the answer is simple: 'by extending the laboratory itself [...] The vaccination can work only on the condition that the farm chosen in the village of Pouilly-le-Fort for the field trial be in some crucial respects transformed according to the prescriptions of Pasteur's laboratory'. After a series of negotiations the scientists persuade the farmers involved in the trial of the need for disinfection, cleanliness, conservation, timing, recording and so forth. Thus, as the trial unfolds, any clear distinction between the laboratory and the farm begins to breakdown: as Latour (1983: 154) puts it: 'no one can say where the laboratory is and where society is'. This result emerges as the laboratory, first, reproduces inside its walls an event that was happening outside – the spread of anthrax – and, second, extends to all farms something that had previously happened only inside the laboratory – disease prevention through vaccination.

The extension of the laboratory into the wider society is given a huge impetus once the field trials are declared successful. At this point, a new fact gains wide acceptance, especially amongst the farming community. This fact is summarized by Latour (1983: 152) in the following way: 'If you want to save your animals from anthrax, order a vaccine flask from Pasteur's laboratory, Ecole Normale

Superieure, rue D'Ulm, Paris. In other words, on the condition that you respect a limited set of laboratory practices [...] you can extend to every French farm a laboratory product made in Pasteur's lab'. Thus, as the vaccine spreads so do the laboratory conditions. In the process many farms are transformed. But more than this, Pasteur transforms French society: he modifies the forces that make up this society and stirs in some new entities – microbes. In this way, Latour argues, Pasteur endows himself with a fresh and novel source of power:

> Who can imagine being the representative of a crowd of invisible, dangerous forces able to strike anywhere and to make a shambles of the present state of society, forces by which he is by definition the only credible interpreter and which only he can control? Everywhere Pasteurian laboratories were established as the only agency able to kill the dangerous actors that were until then perverting efforts to make beer, vinegar, perform surgery, to give birth, to milk a cow, to keep a regiment healthy and so on. (1983: 158).

Thus, Pasteurian laboratories come, not only to hold the solutions to many of society's ills, but also to change the composition of society itself. Society is remade, for now existing relationships must make room for microbes; and, in making room for microbes, society must also make room for the microbes' legitimate spokesperson – Pasteur. It is for this reason that Latour in a later work refers to this process of laboratory extension as the 'Pasteurization of France' (Latour, 1988).

BOX 3.1

The case of Pasteur shows Latour that:

- Scientists become 'great' and 'powerful' because they are able to enrol allies and to build networks.
- These networks must extend backwards and forwards from scientific centres (such as laboratories) to 'non-scientific' locations (such as farms).
- Thus, networks run across or through space and act to bind situated actors together so the composition of space and the facilitation of action are closely combined.
- The networks are 'heterogeneous': they are made of differing entities and resources. These entities and resources are combined in ways that facilitate the spread of scientific facts and artefacts.

In his study of Pasteur, Latour addresses the question of how laboratories gain their power in the world. He shows that this power emanates from an ability to tie together actors situated beyond the laboratory into networks that enable scientific facts and artefacts to travel far and wide. Latour thus builds upon the laboratory studies, taking them beyond the micro-locale of the lab to the transformations that are wrought on the world at large. In so doing, he criticizes sociologists for their reliance on 'dualisms' such as science/society, micro/macro, content/context, inside/outside. He claims that we can only understand how modern science moves through the world if we leave dualistic modes of explanation behind and concentrate on following scientific actors as they tie other actors into networks. As the networks are consolidated, scientific facts and artefacts can spread outside the laboratories in conditions which ensure their proper functioning: 'there is no outside of science but there are long, narrow networks that make possible the circulation of scientific facts' (Latour, 1987: 167). If the networks function correctly, and if all the enrolled entities remain faithful bearers of the facts and artefacts, then authority flows back up the network to the scientist: she or he comes to be seen as the 'actor', the 'cause' of the network effects. In a similar fashion, Latour claims, Pasteur becomes 'powerful' and all those faithful (natural and social) allies that have contributed to his 'power' simply disappear behind his 'greatness'.

From actors to networks

Latour's study of Pasteur evidently follows from Foucault's ideas about the immanent and ubiquitous nature of power relations. Effectively, Latour adopts a Foucaultian perspective on the 'microphysics' of power in science, and shows how the generation of scientific knowledge relies upon the construction of complex alliances or networks. Importantly, power is seen to lie not in the properties or abilities of the scientists themselves but in the relationships they manage to establish between actors and entities of various kinds (that is, bacilli, vaccines, field trials and farmers). Power thus emerges from *within* the network; it is not something imposed upon it from the outside (Latour, 1986).[2] This perspective on power accords closely with Foucault's later ideas on the productive properties of power relations.

In order to tie together the normative and productive aspects of power, Latour introduces the notion of *translation*, an idea that suggests that if scientific networks are to be extended through space and time, then actors of differing (natural *and* social) types must be 'interested' into the network – that is, their goals must somehow be aligned with those of the scientists. Network alignments, as the case of Pasteur indicates, require some degree of 'normalization' so that productive activities can be effectively co-ordinated; in order to produce a vaccine, natural entities must be regulated and farmers must be disciplined. In later work, Latour goes on to consider in some detail how this process of translation tends to operate. First, he distinguishes two main meanings of the term: 'In the geometric sense of translation it means that whatever you do, wherever you go, you *have* to pass through the contenders' position and to help them further their interests. In the linguistic sense of the word translation, it means that one version translates every other, acquiring a sort of hegemony: whatever you want, you want this as well' (Latour, 1987: 120–1). In both senses translation refers to the ways in which one actor gains the ability to speak *for* another. As Callon and Latour explain it:

> By translation we understand all the negotiations, intrigues, calculations, acts of persuasion and violence, thanks to which an actor or force takes, or causes to be conferred on itself, authority to speak on behalf of another actor or force: 'Our interests are the same', 'do what I want', 'you cannot succeed without going through me'. Whenever an actor speaks of 'us', s/he is translating other actors into a single will, of which s/he becomes spirit and spokesman. S/he begins to act for several, no longer for one alone. S/he becomes stronger. S/he grows. (1981: 279)

For Latour, the social scientific notion of 'interest' is important in understanding the mechanics of translation. Interests lie *between* actors, 'thus creating a tension that will make actors select only, what, in their eyes, helps them reach [their] goals' (Latour, 1987: 121). For translation to be successful, there must be a convergence of interests between actors (what Latour calls 'riding piggyback'), and all interests and interpretations of interests must be channelled into the network and must flow down the network in ways that solidify its shape. In other words, if Pasteur is to build associations between elements, translations must be effected so that they all converge on the same purpose or activity – that is, the refinement and dissemination of a vaccine.

By discussing translation in this fashion, Latour suggests that the successful construction and stabilization of scientific networks requires the building of a *consensus* between the participants. In other words, power relations cannot just be imposed but must be agreed upon. In this regard, Pasteur appears to have been a successful translator of interests, for he

> not only recruited many sources of support, but also strove to maintain his laboratory as the source of the general movement that was made up of many scientists, officials, engineers, and firms. Although he had to accept their views and follow their moves – so as to extend his lab – he also had to fight so that they all appeared as simply 'applying' his ideas and following his lead. These two movements must be carefully distinguished because, although they are complementary for a successful strategy, they lead in opposite directions: the recruitment of allies supposes that you go as far and make as many compromises as possible, whereas the attribution of responsibility requires you limit the number of actors as much as possible. (Latour, 1987: 118–19)

The process of translation allows groups of actors and entities to be assembled within a common endeavour and the greater the number assembled, the greater the influence of the network. However, once the multitude has been drawn together some means of maintaining the associations must be established:

> Pasteur had been able to convince farmers who raised cattle that the only way to solve the terrible anthrax plague was to pass through his laboratories at the Ecole Normale

Superieure in Rue d'Ulm in Paris. Breathing down Pasteur's neck were thousands of
interests nested into one another, all ready to accept his short cut through the microscope,
the artificial culture of microbes, and the promised vaccine. However there is a considerable drift between an interest in raising cattle on a farm and watching microbes
grow in petri dishes: the gathering crowd might disband easily. After a few months of
hope they might all leave disappointed, bitterly accusing Pasteur of having fooled them
by creating artefacts in his laboratory of little relevance to farms and cattle. Pasteur
would then become a mere precursor for the anthrax vaccine, his role in history being
accordingly diminished. (Latour, 1987: 122)

Thus, once an initial translation has been achieved, something more is required
to turn the network into a durable whole.

Is there anything that can be used to tie in the farmers' interests before they all go
away bitter and scornful? A tiny bacillus inside a urine medium will not do, even if
it is visible under the microscope. It is only of marginal interest to people who have
been attracted to the lab by the promise that they will soon be back on their farms,
milking healthier cows and shearing healthier sheep. If Pasteur was using his bacillus
to do biochemistry or taxonomy, deciding if it was an animal or a lichen, others like
biochemists or taxonomists would be interested, but not farmers. When Pasteur
shows that sheep fed older cultures of the bacillus resist the disease even when they
are later fed virulent cultures, biochemists and taxonomists are only casually interested but farmers are very interested. Instead of losing interest they gain it. This is a
vaccine to prevent infection, something easy to relate to farm conditions. But what
if the vaccine works erratically? Again, interest may slacken and disappointment
returns. Pasteur then needs a reliable method to turn the production of vaccine into
a routine, a black box that may be injected by any vet. His collaborators discover that
it all depends on the temperatures of the culture: 44 degrees for a few days is fine,
the culture ages and may be used as a vaccine; at 45 degrees, the bacillus dies; at
41 degrees it changes form, sporulates and becomes a vaccine. These little details are
what clamp together the wavering interests of the enrolled farmers. Pasteur has to
find ways to make *both* the farmers and the bacillus predictable. And he has to keep
on discovering new ways, or at least for as long as he wishes to tie these farmers and
these microbes together. The tiniest loose end in this lash up and all his efforts are
wasted' (Latour, 1987: 123–4)

The network thus solidifies around the bacillus, the vaccine and Pasteur's ability
to disseminate the vaccine in ways that effectively inoculate farm animals against
anthrax. As noted above, successful inoculation requires the spread of 'lab conditions' as farmers are encouraged to adopt practices that allow the vaccine to
work *in situ*. In Latour's terms, this means the network must retain an ability to
transform space; as long as spaces are transformed (that is, farms are 'cleaned up')
then the network can be extended; if spaces are not transformed (that is, farms
remain 'dirty') then the vaccine will fail, leading to the eventual breakdown of
the Pasteurian network.

With the network consolidated, Pasteur's lab in Paris effectively becomes
what Latour calls a 'centre of calculation', that is, a discrete space able to act
effectively on many other dispersed spaces. It can act at a distance as long as the

network – the links established between bacillus, vaccine and farmer – remain in place. And in the consolidation of the network, material artefacts play a key role for they, in effect, become 'delegates', able to carry 'rationalities of rule' generated by the centre out to all the localities enrolled in the network. However, these delegates – termed 'immutable mobiles' (Latour, 1987) – must do more than this: they must also carry aspects of the enrolled localities back to the centre. And they must undertake this task in such a way that when the centre holds the mobiles it also holds some very real facets of the localities themselves. Only in this way can control over the network be achieved and maintained.

Latour thus defines some general features of centres of calculation. First, they must somehow 'bring home' relevant features of the places and peoples of concern. This can be done by (a) rendering them mobile so that they can be moved, (b) keeping them stable so that they retain their shape, and (c) making them combinable so that 'whatever stuff they are made of can be cumulated, aggregated, or shuffled like a pack of cards' (1987: 223). In the case of Pasteur, these features apply to the anthrax bacillus, which is refined in the field and is then transported back to Paris to be combined with other elements so that eventually a vaccine can be produced. The second stage is for the centre to reach back out to the multitudes of micro-locales upon which it might act. Again, the ability to transport stable elements back out into the world is the crucial issue. As the case of Pasteur shows, the stabilized element (the vaccine) requires that conditions outside the centre are made propitious for its functioning. If the network is successfully extended, Pasteur's facts and artefacts can flow outward into French agriculture and French society.

There was nothing more dramatic at the time than the prediction solemnly made a month in advance by Pasteur that on 2 June 1881 all the non-vaccinated sheep of a farm in the little village of Pouilly-le-Fort would have died of the terrible anthrax disease and that all the vaccinated ones would be in perfect health. Is this not a miracle, as if Pasteur had travelled in time, and in the vast world outside, anticipating a month in advance what will happen in a tiny farm in Beauce? If, instead of gaping at this miracle, we look at how a network is extended, sure enough we find a fascinating negotiation between Pasteur and the farmers' representatives on how to transform the farm into a laboratory. Pasteur and his collaborators had already done this trial several times inside their lab, reversing the balance of forces between man and diseases, creating artificial epizootics in their lab. Still, they had never done it in full-scale farm conditions. But they are not fools, they know that in a dirty farm thronged by hundreds of onlookers they will be unable to repeat exactly the situation that had been so favourable to them [...] On the other hand, if they ask people to come to *their* lab no one will be convinced. [So] they have to strike a compromise with the organisers of a field test, to transform enough features of the farm into laboratory-like conditions – so that the same balance of forces can be maintained – but taking enough risk – so that the test is realistic enough to count as a trial done outside. (Latour, 1987: 248)

BOX 3.2

In Latour's view building networks requires:

- Processes of 'translation' must be executed so that actors and entities are enrolled into network relations.
- 'Translation' means that the enrolled actor is persuaded to 'identify' with the network. This may mean some modification in the actor's identity and/or it may mean some modification in the shape of the network to accommodate a new actor.
- 'Translation' can be executed either consensually or coercively, or through some combination of the two. Actors can be persuaded to join the network because they come to believe it is in their 'interests', or they can be forced to join against their 'interests'.
- Once enrolled into the network, the relations between entities must be stabilized. These stabilizations are often delegated to non-human entities such as technologies, because materials of various kinds are themselves generally more stable than human actions. In short, technologies can make good disciplinary machines.

Associational action

Latour's analysis of Pasteur's anthrax vaccine illustrates how science works to 'colonize' a range of locales beyond the laboratory. Using this case study, he shows that science only 'works' if scientists somehow 'change the world' in ways that correspond with conditions inside the lab (see also Rouse, 1987). The influence of the laboratory on the outside therefore works in two ways. First, elements of the outside world (in this case the anthrax bacillus) are brought into the lab to be analysed and altered. Second, the modified elements are exported back out into the world in order to effect change of some kind. Both these influences require networks, defined as heterogeneous associations of actors and entities. The networks allow elements to flow towards the centre of calculation (the lab) and then back out again into a host of micro-locales.

We can see, then, that laboratories gain their powers from the associations they bring into being. They can act over long distances but any actions they undertake have to be conducted *through* the many other actors and entities that have been enrolled into the networks. Thus, in actor-network theory action – as the case of Pasteur clearly shows – arises from collective endeavour and the collective includes both humans and non-humans. As Latour (1999: 192) puts it: 'Purposeful action and intentionality may not be the properties of objects, but they are not properties of humans either. They are the properties of institutions, of apparatuses, of what Foucault calls dispotifs'. This relational view of action follows from the idea that 'nature' and 'society' are co-constructed in the laboratory. In the experiments conducted by scientists such as Pasteur, Latour (1999: 228) notices that action comprises 'not what people do' but 'what is accomplished along with others'. Action is therefore the result of network mobilization and networks rely on entities of many kinds.

In many ways, Latour is adopting a realist approach here as, in his view, 'things' play an effective role in social life – that is, they are more than just 'social constructions'. However, he also adopts a constructivist form of realism for he believes things only take shape in networks. In his study of Pasteur, he is interested in the anthrax bacillus only once it emerges as a discrete and autonomous entity in Pasteur's experiments. Thus, Latour argues, we should not imagine the bacillus is simply a thing 'out there' waiting to be discovered by intrepid humans 'in here' (or, for that matter, that the thing 'out there' is simply constructed by the human 'in here'). It is the *co-construction* of a complex socio-natural assemblage that allows the (natural) substance (and also the 'great scientist') to emerge. Thus, 'when a phenomenon "definitely" exists that does not mean that it exists forever, or independently of all practice and discipline, but that it has been entrenched' in a network (Latour, 1999: 155–6). And in a network, all entities are assembled 'symmetrically': that is, the 'natural' entities are just as likely to be active as those labelled 'social', so that processes of 'construction' cannot be seen as emanating from purely social or human causes.

Latour's colleague Michel Callon (1986) provides a clear illustration of action arising from the combined relations of humans and non-humans when he examines the application of scientific knowledge to scallop fishing in northern France. In a revealing (and much-cited) case study, Callon tells how a group of scientists attempt to persuade a group of French fishermen of the utility of their scientific knowledge by specifying a set of guidelines which will increase scallop numbers. Callon shows how the scientists attempt to build a scientific network by getting other actors to comply with them. As the scientists link the entities together, so they designate a set of interrelated roles. Importantly, the entities include non-humans, and Callon shows how the scientists enrol both scallops and fishermen into their network. However, he also goes on to show that for the network to be successfully stabilized, the designated roles have to be accepted by all the actors. In this case, the fishermen and the scallops reject their allocated functions and effectively go their own way, thereby breaking apart the network. As well as showing how processes of network construction can fall apart, this outcome indicates that non-humans can be just as effective in initiating action as humans. There are of course countless examples of non-human action: we might think of BSE where a new actor (a so-called 'prion protein') escaped from one set of relations within the food chain and linked together a new set of associations, incorporating cows, abattoirs, politicians, beefburgers and so forth (see Hinchliffe, 2001). Similarly, the explosion of the nuclear reactor at Chernobyl brought together a new set of associations between reindeer, rain, Cumbrian farmers, grass and scientists (see Wynne, 1996). While it might be argued that these are exceptional cases, they are exceptions that prove the rule: things act in concert with humans; humans act in concert with things.

Both classes of entities are associated within networks and retain the ability to act within network relations.

At this point it is worth pausing to consider how the notion of agency asserted by actor-network theory differs from that prevailing in much mainstream social science, for it seems that a truly relational view of the social actor is being asserted here. Fuller points to a key distinction between the actor-network theory view and traditional perspectives: 'instead of treating agency as an ontological primitive out of which societies are constructed [actor-network theory] treats agency as a theoretical construct carved out of an already transpiring social order' (1994: 746). The 'primitive view', referred to by Fuller, sees the agent as an already formed solid mass moving according to its own principles and tendencies unless impeded by other forces (for instance, power imposed from the outside). In contrast, the actor-network view of agency begins not with fully formed agents but with an already constituted social space (the network) and shows how agents (both human and non-human) emerge from a series of trials in which they are continually striving to become actors with powers (for instance, the relations between anthrax bacilli, vaccines, farmers and French society). It is only at the end of a period of stabilization that the actors can be distinguished from the lesser entities, which by now are simple intermediaries (that is, Pasteur has achieved actor status while all the others – bacilli, vaccines, and farmers – serve merely as linkages in the network).[3] 'Who will win in the end? The one who is able to stabilise a particular state of power relations by associating the largest number of irreversibly linked elements' (Callon and Latour, 1981: 293). Thus, actors are 'effects generated in configurations of different materials' (Callon and Law, 1995: 502), while action is the property of associations rather than agents: 'the prime mover of an action becomes a new, distributed and nested series of practices whose sum might be made but only if we respect the mediating role of all the actants mobilised in the list' (Latour, 1994: 34). Action thus emerges from association and responsibility becomes distributed along the chain of humans and non-humans. As Callon and Law (1995: 485) put it: 'it's the relations [...] that are important. Relations which peform. Perform agency'.

The important point to again note here is that actors and entities only emerge from within – that is, their shapes and forms are only determined by – the relations established in networks. Although, as Latour notes, the subjects and objects incorporated into networks bring pre-existing identities into the here and now ('we hourly encounter hundreds, even thousands, of absent makers who are remote in time and space yet simultaneously present', Latour, 1994: 40), the network does not emerge as a simple aggregation of these already stabilized entities, for all are modified as they enter into new and complex interrelationships. 'External' identities become what Brian Wynne (1996: 362) calls '(inter)dependent variables'. However, these variables do not sit outside the fields of negotiation and construction in which the networks are made but are 'reshaped (and variably stabilised, temporarily) in the [...] heterogeneous processes of co-construction and mutual reinforcement' (1996: 362). This reshaping results from processes of translation.

> This crucial operation engenders the establishment – albeit local and provisional of social links. Thanks to translation, we do not have to begin our analysis by using actants with fixed borders and assigned interests. Instead, we can follow the way in which actant B attributes a fixed border to actant A, the way in which B assigns interests or goals to A, the definition of these borders and goals shared by A and B, and finally the distribution of responsibility between A and B for their joint action. (Latour, 1991: 127)

What the actor-network theorists seek to investigate, then, are the means by which associations between actors and entities come into existence and how the roles and functions of subjects and objects, actors and intermediaries, humans and non-humans are attributed and stabilized. They are interested in how these and other categories emerge from processes of network building. Actor-network theorists make the radical claim that it is only as a result of network-building activities that any stable categories emerge – categories do not exist outside specific network formations. Moreover, an actor (social or otherwise) will only come into being if the links established between the entities enrolled in the networks allow one of their number (perhaps the entity that initiated the enrolment process) legitimately to claim actor status (that is, power flows down the chain towards she/he/it, elevating her/his/its status above all the others). Thus, in an important sense the distinction between actors (those that *organize* the associations or networks) and intermediaries (those that are *organized* within networks) comes at the end of the construction process, when the former can take credit for the latter. However, we should remember that it is only through the (translated) efforts of these associated others that the actor is able to grow in size and extend its reach over greater distances, becoming in effect global: 's/he begins to act for several, no longer for one alone. S/he becomes stronger. S/he grows' (Callon and Latour, 1981: 279).

BOX 3.3

Latour's focus upon actors-in-networks leads to the following assumptions:

- Actors can only act in concert with others. Actors only become actors if those others conduct actions in ways approved and recognized by 'the' actor.
- Action is thus profoundly relational: it can only take place because of the alignments of actors, entities and resources. These alignments are common, everyday features of socio-spatial life.

Continued

- Actors, entities and resources only finally take shape (acquire identity) within network relations (any pre-existing identities are likely to be modified or displaced during the process of enrolment). Actors and entities are therefore co-constructed in networks.
- Because the networks are heterogeneous in nature, then a host of actors and entities must be mobilized to make any action effective. This means that if any actor or entity leaves the network the whole operation is threatened. Thus, all the enrolled entities have 'power' of some kind.
- This view of action means we should adopt a 'symmetrical' perspective on potential actors: both humans and non-humans have the ability to make moves that hold decisive implications for the network as a whole.

We therefore arrive, finally, at the *actor*-network: actors and networks become one and the same: it is now 'all for one and one for all' in the construction of joint actions. And as the actor-network grows, it will extend its influence and reach beyond a single locale into other locales, tying these together in sets of complex associations. There is, therefore, no difference in kind between 'macro' and 'micro' or 'global' and 'local'; in the view of actor-network theory longer networks simply reach further than shorter networks.

Network space

The notion that a laboratory is a centre of calculation, able to act at a distance on many diverse micro-locales, helps us to understand how spatial relations are established between sites. It seems from the above account that an interaction between network and site is required so that the site can be modified in line with the requirements of the network. As Latour puts it:

> Every time a fact is verified and a machine runs, it means that the lab or shop conditions have been extended *in some way* [...] forgetting the extension of the instruments when admiring the smooth running of facts and machines would be like admiring the road system, with all those fast trucks and cars, and overlooking civil engineering, the garages, the mechanics and the spare parts. Facts and machines have no inertia of their own; like kings and armies they cannot travel without their retinues or impedimenta. (1987: 250, original emphasis)

Through translation processes, it is possible to do things in one place (for example, the centre) that dominate another place (for example, the periphery). So the term 'local' has a double meaning: first, it refers to the coordinated practices of actors in some predefined locality (for example, the laboratory); second, it refers to the strategies of 'localization' being employed as places are 'lined up' within a given network.

This concern for strategies of localization allows actor-network theory to move beyond the micro – macro divide that ultimately proved problematic for Foucault. By drawing our attention to networks, the theory suggests that everything takes place at 'ground level'; there is no need to jump between spatial scales. Latour illustrates this point by asking of a railroad, 'is [it] local or global?' The answer he provides is neither, for

> it is local at all points, since you always find sleepers and railroad workers, and you have stations and automatic ticket machines scattered along the way. Yet it is global, since it takes you from Madrid to Berlin or from Brest to Vladivostock. However, it is not universal enough to be able to take you just anywhere. It is impossible to reach the little Aubergnat village of Malpy by train, or the little Staffordshire village of Market Drayton. There are continuous paths that lead from the local to the global, from the circumstantial to the universal, from the contingent to the necessary, only as long as the branch lines are paid for. (1993: 117)

Thus, 'the words "local" and "global" offer points of view on networks that are by nature neither local nor global but are more or less long and more or less connected' (1993: 122). Size and scale are nothing more than the end product of network extension processes. For actor-network theorists, then, geographical analysis means staying within the networks: we should never vacate the local to look for explanations at another scale of analysis. Yet, neither should we remain trapped in the local, for this spatial demarcation only makes sense in the context of larger network formations. We should travel from locale to locale paying particular attention to the various relationships that serve to bind places together: 'though places are distant, irreducible, and summable, they are nevertheless constantly brought together, united, added up, aligned and subjected to ways and means. If it were not for these ways and means, no place would lead to any other' (Latour, 1988: 164).

To understand the construction and consolidation of space and time, we must therefore follow the networks wherever they might lead. To do this the actor-network theorists believe we must follow a simple methodology:

> We have to be as undecided as possible on which elements will be tied together; on when they will start to have a common fate, on which interest will eventually win out over which. In other words, we have to be as undecided as the actors we follow [...] The question for us, as well as for those we follow, is only this: which of the links will hold and which will break apart? (Latour, 1987: 175–6)

Importantly, this act of 'following' requires that we do not specify different levels of analysis in advance. Callon et al. believe the adoption of a single framework is crucial if we are to grasp the establishment of 'equivalences between places':

> to make use of a separate vocabulary for the large tends to conceal both the processes by which growth occurs, and the uncertainties that are involved in maintaining power and size. In addition it reifies the status of the large, and makes it appear as if the latter

could never be decreased in size and become weak. We believe the social theorist has no reason to aid those who happen to be powerful. (1986: 228)

Actor–network theorists thus reject the view that social life is arranged into levels or tiers (some of which determine what goes on in others).[4] It is only the mobilization of humans and non-humans across space and time that distinguishes the 'local' from the 'global', the 'macro' from the 'micro'.

> Instead of having to choose between the local and the global view, the notion of network allows us to think of a global entity – a highly connected one – which remains nevertheless continuously local [...] Instead of opposing the individual level to the mass, or the agency to the structure, we simply follow how a given element becomes strategic through the number of connections it commands and how it loses its importance when it loses connections. (Latour, 1997a: 3)

This 'grounded' approach suggests that length of network determines scale – some networks remain tied to what we would normally see as local areas, other extend over distances we might term 'national', while yet others run around the world in a 'global' configuration. The networks stabilize spatial relations using a range of resources, assembled in ways that allow the flow of knowledge, materials, personnel and so forth up and down the network from the centre outwards. The transported entities

> travel inside narrow and fragile networks, resembling the galleries termites build to link their nests to their feeding sites. Inside these networks, they make traces of all sorts circulate better by increasing their mobility, their speed, their reliability, their ability to combine with one another. We also know these networks are not built with homogeneous material but, on the contrary, necessitate the weaving together of a multitude of different elements which renders the question of whether they are 'scientific' or 'technical' or 'economic' or 'political' meaningless. Finally, we know that the results of building, extending and keeping up these networks is to act at a distance, that is to do things in the centres that sometimes make it possible to dominate spatially as well as chronologically the periphery. (Latour, 1987: 232)

The combination of materials used to build the networks will vary in accordance with the types of relations to be consolidated. In the case of Pasteur, the aim is to tie the laboratory in Paris to the many farms in France where anthrax is an acute problem. The linkages therefore have to be formed around the solution to the anthrax problem: the refinement of the bacillus, the development of a workable vaccine and the modification of farm conditions. But more than this, the linkages have to convey the vaccine into the outside world while *at the same* time conveying power, responsibility and recognition back towards the lab and ultimately to Pasteur. In other words, Pasteur's lab has to become an 'obligatory passage point' for all solutions to the anthrax problem (Latour, 1987). The choice of materials therefore reflects this overall goal, the need to engineer this two-way movement. And in forging this network in which entities flow in and out of the lab, Pasteur creates a new space-time configuration where his lab is extended to almost every farm in France. This new spatial formation has its global and its local aspects: Pasteur's laboratory is 'globalized' through its affiliation with many French farms, while the farms are 'localized' by the nature of their relations with the lab.

BOX 3.4

Notions of space in Latourian actor-network theory:

- Space (and time) are constructed within networks; they are 'made' out of relations of various kinds.
- Thus, in order to analyse particular spaces (and times) we must 'follow' the networks in order to follow the processes that construct space (and time).
- The networks never shift registers or scales. So, in following the construction of space (and time), we never need to shift from the 'micro' to the 'macro' or from the local to the global; rather we just follow the networks wherever they might lead.
- Actor-network theory therefore provides a single terminology (and a single methodology) for the study of space and spatial relations. It simply emphasizes the need to follow networks and to study the materials they are made of and the relations established between these materials.

Actor-network theory therefore proposes a firmly relational view of space. Networks create various space-times out of the materials they bring together ('each invention of a new immutable mobile is going to trace a different space-time' Latour, 1987: 230). These space-times are extended as the networks draw external locales within their spheres of operation. In the case of Pasteur, the network draws in farms by arranging an interaction between the lab in Paris and conditions on the farm. Once this linkage is established, the bacillus can flow towards the lab and the vaccine can flow towards the farms. The space of the network is conditioned by the need to ensure this two-way flow.[5] In sum, 'space' is nothing more than a network 'effect'. Latour puts the point starkly when he says:

we should force [the] immense extents of space and time generated by geology, astronomy, microscopy, etc., back inside their networks – these phentograms, billions of electrovolts, absolute zeros and eons of times; no matter how infinitely big, long or small they are, these scales are never much bigger than the few metre squares of a geological or an astronomical map, and never much more difficult to read than a watch. We, the readers, do not lie *inside* space, that has billions of galaxies in it; on the contrary, this space is generated *inside* the observatory by having, for instance, a computer count little dots on a photographic plate. To suppose, for example, that it is possible to draw together in a synthesis the times of astronomy, geology, biology, primatology and anthropology has about as much meaning as making a synthesis between the pipes or cables of water, gas, electricity, telephone and television. (1987: 229)

There is no absolute space (just as there is no absolute nature, no absolute society, no absolute time); only specific space-time configurations, conditioned by the rationalities and relations that run through networks.

Conclusion

It seems clear that actor-network theory's distinctive perspective on the natural sciences derives in large part from Foucault's earlier approach to the human sciences. In particular, actor-network theory builds upon two of Foucault's crucial insights into the operation of power relations: first, that power is every-where (even in laboratories, those citadels of scientific rationality); second, that power is productive (within laboratories, power relations assist in the genera-tion of scientific knowledge). Actor-network theory thus re-describes scientific activity in a language that somehow captures its suffusion by power relations. For scientists to be successful, they must 'translate interests'; they must build networks on terms that allow power and authority to flow towards them; they must extend the networks by extending laboratory conditions; they must deliver new definitions of both nature and society; they must, in John Law's (1986) phrase, become 'heterogeneous engineers'.

Actor-network theory also extends Foucault's approach by focusing upon the variety of materials that allow power to flow up and down the networks. As Allen (2003: 131) says, Latour 'has helped to render visible something of what is involved in establishing and maintaining ordered lines of conduct at a distance'. In many respects it is the combining of materials in durable and effective formations that allows conduct to be ordered in this way. Actor-network theory makes the bold claim that it is only through heterogeneous networks that actors make any impact upon the world; no actor can make any kind of effective intervention without the support of others; action is associa-tion. Thus, the associated actor is an actor-network, and the actor-network is a stable, enduring and effective ensemble of actors and entities, combined in ways that allow a centre to gather resources in and to export its products out.

Actor-networks can be found in science but in other domains as well, for the processes involved in making scientific knowledge – translation, enrolment, network extension – apply in almost all areas of social life. In short, there is no such thing as society: only (heterogeneous) networks. And space, too, is made ('materialized') by these networks. The alignment of heterogeneous resources within networks leads to the fabrication of various space-time configurations. Spaces emerge from the weaving together of localities in line with the precepts of the network. Thus, localities are 'localized' according to the rationalities and practices that make the network what it is. In most cases, processes of localization will occur as network 'nodes' work to establish durable structures of centralization and peripheralization within the networks; as these structures are constructed so discrete spaces are marked off from one another in the context of a network 'hierarchy' of centres and peripheries.

This brings us back to the relationality of space. In actor–network theory space is no longer absolute (something networks exist *within*); rather, space is an effect of network activity. It emerges from within heterogeneous networks and its shape and its form is given by the shape and form of the various networks. As Latour (1988: 25) puts it: 'Gods, angels, spheres, doves, plants, steam engines, are not *in* space and do not age *in* time. On the contrary, spaces and times are traced by reversible or irreversible displacements of many types of mobiles. They are generated by the movement of mobiles, they do not frame these movements'. Networks and the entities that flow through them *make* space; thus, multiple networks make *multiple* spaces. There may be some commonality in the delineation of network spaces but more likely there will be discrepancies, discrepancies derived from the differences in make-up. Differing networks co-exist; thus, differing space-times co-exist.

As Bingham and Thrift (2000: 290) point out, actor–network theory is less concerned with space and time than with unique acts of 'timing and 'spacing', acts that are conducted through associations of various kinds. Geography becomes then the study of associations or networks. However, the question this raises is whether any geographical overview of associational or network spaces is possible. Or put another way, can we gain understandings of spatial relations only from *within* networks, or should we somehow 'step outside' given network relations in order to gain some (objective?) understanding of the cumulative effects of multiple networks? From the preceding pages it might be inferred that actor–network theory would see the attempt to gain some general spatial overview as merely the misplaced ambition of a discipline such as geography that has always used 'absolutist' notions of space in order to gain dominion over the spatial realm. Latour, for one, clearly believes such ambitions to be misplaced. As he puts it:

> the difficulty we have in defining all associations in terms of networks is due to the prevalence of geography. It seems obvious that we can oppose proximity and connections. However, geographical proximity is the result of a science, geography, of a profession, geographers, of a practice, mapping, measuring, triangulating. Their definition of proximity and distance is useless for [actor–network theory]. The notion of network helps to lift the tyranny of geographers in defining space and offers us a notion which is neither social nor 'real' space, but association. (1997a: 2)

Actor–network theory therefore poses a challenge to geography: it demands not only that a relational view of space is adopted but also that spatial relations are seen as network relations. Within the network domain, spatial scale is reconceptualized as network length and network length is reconceptualized as 'heterogeneous engineering' – that is, processes of network building in which entities of various kinds are assembled in ways that allow networks to undertake certain functions. Actor-network theory claims that geographical analysis (like all other forms of analysis) should now come down to a few methodological points: follow the actors as they stitch networks together, observe what is linked to what, and assess how power flows up and down from centres to peripheries and back again (Murdoch, 1997). But, in concluding this chapter, we might legitimately ask whether such a simplified methodology is really sufficient grounds for geographical analysis (Thrift, 1999). Can geography simply be recast as network analysis? Must we always stay within single networks or can we make some effort to see beyond particular network arrangements to broader network formations in which multiple networks give rise to multiple space-times that somehow mesh together into a broader spatial context? We take up these questions in the next chapter as we assess more fully the spatial implications of actor-network and Foucaultian theory.

SUMMARY

We have examined Latour's 'version of actor-network theory' here in order to show how geographical locations are aligned in relations of various kinds. The chapter focused mainly on Latour's analysis of Pasteur and showed how a scientific network came into being around this scientist. The network was made of heterogeneous materials – anything that rendered it durable – and connected a range of differing locations. It also mixed up the 'macro' and the 'micro', the 'local' and the 'global'. The chapter therefore examined how scale is made relationally.

FURTHER READING

Latour's work is generally accessible. Probably the best introduction is his (2005) book, *Reassembling the Social*. *We Have Never Been Modern* (1993) is also very readable. For an analysis that teases out the implications of Latour's approach for geography see Sarah Whatmore's (2002) book, *Hybrid Geographies*, and for some criticisms of the actor-network view on power, see John Allen's (2003) book, *Lost Geographies of Power*.

Notes

1. As Cole (1992: 30) says, it was widely assumed that 'natural scientists were trying to discover the next page of a book that had already been written, whose conclusion, though currently unknown, was pre-determined or inevitable. Nature, rather than sociological processes, determined the way in which scientific knowledge developed'.

2. Latour seems to retain something of an ambivalent attitude to Foucault. He professes to like Foucault's (1979) account of power in *Discipline and Punish*, and he believes this provides a model on which empirical studies of network building can be based. However, he expresses reservations about other aspects of Foucault's work, particularly the latter's neglect of the natural sciences, which Latour believes limits the utility of Foucaultian vocabularies and concepts (see the discussion in Latour and Crawford, 1993).

3. Callon and Latour (1981: 286) ask: 'What is an actor? Any element which bends space around itself, makes other elements dependent upon itself and translates their will into a language of its own. An actor makes changes in the set of elements and concepts habitually used to describe social and natural worlds'.

4. Moreover, as Callon (1986: 228) puts it: 'to make use of a separate vocabulary for the large tends to conceal both the processes by which growth occurs, and the uncertainties that are involved in maintaining power and size. In addition it reifies the status of the large, and makes it appear as if the latter could never decrease in size'. In order to better understand the precarious nature of network building, we should attend to the processes involved; these can only be seen from within the network itself rather than from the perspective of another spatial scale.

5. Space can also 'reappear' if the flow is for some reason interrupted. An illustration of this point is provided by Latour (1997b: 173), when he cites the example of a passenger on board a TGV train: 'He sat quietly in the first class, air-conditioned passenger car and read his newspaper, paying no attention to the many places passed by the speeding train, all of which looked to him like landscapes projected on a movie screen [...] No negotiation along the way, no event, hence no memory of anything worth mentioning'. This entity, this traveller, is transported relatively unchanged across space (he almost 'hovers above' the places passed en route) along a given time horizon. But if the train breaks down and the passengers are forced to disembark then they suddenly become very concerned about space. This space is no longer a 'landscape on a movie screen', a mere passing facade – something rendered almost invisible by the compression of time – but is now a complex and concrete place which the passengers have to negotiate during the time of their delay.

4

Space in a network topology

To attend to the politics of becoming is to modify the cultural balance between being and becoming without attempting the impossible, self-defeating task of dissolving solid formations altogether. (Connelly, 1999)

Introduction

Actor-network theory builds on Foucaultian theory by showing how power is conducted within network formations. Power, in this view, lies in the heterogeneous materials assembled in networks in accordance with the need to make actions (scientific or otherwise) durable through space and time. Networks draw materials together into new configurations. Each network traces its own trajectory and this trajectory reflects a convergence of factors, including the combination of entities used in network construction, the relations established between these entities, and the ordering impulses of the network builder. If all these elements work in concert then the network becomes a solidified actor – an 'actor-network'. This term, which Law (1999: 3) claims is deliberately 'oxymoronic', refers, on the one hand, to a centred actor and, on the other, to a decentred network. Actor-networks are networks *and* points, individuals *and* collectives (Callon and Law, 1997: 174).

Latour and the other actor-network theorists believe it is the mixing of human actions into non-human materials which allows networks to endure beyond the present and to remain stable across space. It is therefore the heterogeneous quality of the networks that permits them to reach across spatial scales from the 'small scale' to the 'large scale'. Actor-network theory thus directs our attention to the means whereby spaces are made ('materialized') *inside* networks and it shows how spatial scales are distinguished from one another in line with the priorities of the networks or the network builders. In this respect, we can suggest actor-network theory extends Foucault's analysis of 'Panopticism' for, as we saw in Bentham's Panoptic prison, certain centrally placed actors (the guards) could 'localize' the prisoners by rendering them visible. The architecture of the prison was determined by this requirement for 'localization' and 'visualization'. As the prisoners were confined and observed so their behaviour was 'normalized'. Thus, the arrangement of materials in the prison corresponded to the requirements of a specific disciplinary regime.

In Latour's analysis of Pasteur, we see precisely the same mechanisms at work. Pasteur turns his laboratory into a centre of calculation by rendering the anthrax bacillus visible. Once this visibility is achieved, the bacillus can be acted upon until a vaccine is developed. As the vaccine is exported, laboratory conditions are extended until it is no longer clear where the lab ends and the external environment (that is, French society) begins. In Latour's (1983, 1987, 1988) account, it appears that Pasteur is a network builder *par excellence*: he is able to align a host of entities in a way that permits the extension of his laboratory, while at the same time ensuring recognition and prestige flow towards himself. In short, Pasteur constructs an actor-network in which he is ultimately seen as 'the actor'. His laboratory in Paris becomes a centre of calculation, while all other spaces are positioned as somehow peripheral (despite their importance to the network as a whole). From this, we might assume that the networks in actor-network theory refer to systems of almost Panoptic power in which centres succeed in exercising effective control over all aligned entities and spaces.

Yet, subsequent studies have shown that Panoptic networks are not necessarily the norm; they may co-exist with much more fluid network relations, perhaps echoing Foucault's point that systems of domination comprise only one form of power relation. We can therefore suggest the existence of two broad network types (Murdoch, 1998). On the one hand, there are those networks where translations are perfectly accomplished, where the entities are effectively aligned and the network is stabilized – despite the heterogeneous character of the entities they work in unison, thereby enabling the enrolling entity (the 'centre') to 'speak for' the entire network (rather in the manner of the Panoptic prison and the Pasteurian laboratory). On the other hand, there are networks where the links between actors and intermediaries are provisional and divergent, where norms are hard to establish and standards are frequently compromised. Here the various components of the network continually negotiate with one another, forming variable and revisable coalitions, and assuming ever-changing shapes so that no clear centre emerges. While this second type might be seen as an early version of the first – once relations are settled then a dominating centre will emerge and norms will be imposed – it does not always work out this way; sometimes networks take shape in non-centred ways (Callon, 1992).

It seems reasonable to assume that these two network types demarcate differing spaces: in 'Panoptic' networks, spaces are strongly prescribed as delegates, mobiles, inscriptions and other envoys work to 'normalize' behaviour; in networks of variation and flux, alignments are interactional and unstable, giving space a more malleable character. These two network spaces might be described as 'spaces of prescription' and 'spaces of negotiation' (Murdoch, 1998). These terms refer to the degree to which networks are controlled by powerful centres or whether they form out of only loose and shifting affiliations. In certain networks, centres are able to prescriptively line up entities in ways that enhance their control; in other networks, discretion and negotiation between all the assembled elements are the norm.

There is, then, more than just one set of spatial relations in the scenarios portrayed by actor-network theory. On the one hand, we have tightly ordered ('normalized') spaces; on the other hand, we have disordered ('undisciplined') spaces. As we shall see in this chapter, the distinction between 'prescription' and 'negotiation' is helpful in highlighting the differing sets of relations consolidated in networks. Moreover, it directs our attention to the varied sets of heterogeneous associations that compose differing spatial arenas. Yet, the two-fold typology 'prescription/negotiation' may be unduly restrictive when we turn to examine the relationships between differing networks and differing spaces. Thus, we introduce to the notion of 'multiplicity' as perhaps a better means of describing spatial complexity. The term multiplicity helps us to appreciate undulating landscapes of network relations in which differing spatial contacts coexist. An interest in multiplicity therefore leads directly to a concern for 'topology' – that is, the complex spatial interactions that take place both within and between networks. In what follows, we investigate the utility of these two concepts – multiplicity and topology – and assess how far each helps us to appreciate the interaction between network and space.

In assessing the notions of multiplicity and topology, we also address some of the challenges that actor-network theory poses for the practice of human geography. In particular, we consider whether *general* observations about (relational) space can be made from within a framework that suggests there are as many spaces (and times) as there are networks. In other words, given that actor-network theory appears to question whether 'geography' can legitimately gain access to any vantage point that provides an *overview* of the spatial realm, we assess whether geographers can still legitimately stand above 'ground level' in order to survey the broader socio-spatial terrain.

In order to address these issues, the chapter is divided into two main parts. In the first, we assess the way in which Foucaultian/actor-network theory conceptualizes network space. As we shall see, recent thinking in this theoretical stream highlights spatial complexity, for it is now evident that networks give rise to differing spatial forms. We consider how spatial complexity leads to an interest in multiplicity and topology, and we assess how Foucaultian/actor-network theory has attempted to incorporate these terms into its repertoire. In the second part of the chapter we move beyond the discussion of network space to consider the broad landscapes of spatial relations that arise in this 'world of networks'. In particular, we assess whether it is possible to develop a secure philosophical vantage point for broad geographies of relational space. In seeking a vantage point somewhere above 'ground level', we turn to the work of Gilles Deleuze and Michel Serres – two key philosophical influences on actor-network theory. Through a discussion of the relational geographies that emerge from their thinking we begin to assess how some generalized geographical perspectives might be developed in line with the relational requirements of

Foucaultian/actor-network theory. In conclusion we begin to draw out an analytical framework that combines both topographical and topological perspectives. This framework will guide the discussions presented in the case-study chapters.

Spaces of singularity and multiplicity

As we observed in the previous chapter, actor-network theory has been particularly adept at showing how action at a distance is achieved through the use of heterogeneous materials. 'Translation' is the conceptual tool most frequently utilized in the theory as an aid to this type of investigation. In short, translation refers to the processes of negotiation, mobilization and displacement that aim to establish enduring relations between actors, entities and places. It involves the re-definition of these phenomena so that they are persuaded to behave in accordance with network requirements and these redefinitions are frequently inscribed in the heterogeneous materials that serve to consolidate the networks. And as we indicated above, there is a close affinity between this approach and Foucault's analysis of normalizing power relations in institutional settings such as prisons and asylums.

Although the term 'translation' focuses our attention on the negotiated character of network enrolments, actor-network theorists sometimes imply that this is a *prescriptive* process. To take just one example, Law (1997: 4) suggests that 'networks may be imagined as scripts. Which means that one may read a script from, for instance, a machine which tells or prescribes the roles that it, the machine, expects other elements to play'. As indicated above, where a network behaves in this way, it is likely to be standardized and predictable. The most predictable networks tend also to be the most formal. In general, formalisms are composed of separate countable elements which stipulate a hierarchy of spatial and temporal relationships. As Bowers explains, these separate, countable elements provide a means of

> manipulating a few elements, combining and recombining them systematically, while practices of re-representation [or translation] retain the link between the few formal elements and many other representations [translations] which stand behind/before them [...] Like the strands in a rope, there are a multiplicity of well-ordered and combined elements connecting one end (the object) with the other end (the formalism). (1992: 245)

Thus, the elements of the network 'fold up' the representatives that stand behind them so that the network becomes 'singular' – it becomes an *actor-* network. If the network is to achieve 'actor' status then entire chains of translation must be arranged into complex hierarchies. These hierarchies will resonate centrally-generated impulses to the extent that each component reflects the whole – despite their heterogeneity, the assembled entities work in unison (Callon, 1991). In other words, the network becomes a singularity.

While actor-network studies show that there are various ways of lining up formal sets of relations 'behind' centrally-placed actors, amongst the most common are those based upon numerical inscriptions. For instance, various commentators have shown that accounting procedures work effectively as a mode of regulating action across space. Robson explicitly uses Latour's notion of 'action at a distance' to explain how accounting works in this regard.

> Accounting records provide a regular form of surveillance. As the actions of the worker are inscribed in costing and production records a basis for constructing productive norms and trends, and work targets is created. Not only is this form of control continuous, it is also impersonal, proceeding in the absence of face to face contact with supervisors and management. By gathering accounting numbers and collating them into divisional, regional or organisational measures, or averaging accounting measures into standards of input or output per employee to which each employee's performance can be referred and evaluated, the accounts create a visibility, bases for calculation and thence an opportunity for acting upon individuals and productive processes through the medium of accounting measures and targets. (Robson, 1992: 700)

We see here that accounts – as 'immutable mobiles' that can travel back and forth between centre and periphery – work to bind various locales into central modes of calculation. These modes of calculation can then prescribe what goes on in the various locales so that behaviour conforms to the rationalities of accounting. We can also see from this quotation that Foucault's concept of 'normalizing judgement' applies fairly accurately to the workings of prescriptive accounting networks. The formalisms that are routinely used in such networks serve to regulate behaviour in ways that seemingly leave little scope for autonomous action on the part of locally-situated actors. Behaviour is 'normalized' in line with network precepts – for example, the need to calculate in accounting terms. Or, to put it another way, techniques such as accounting ensure the 'conduct of conduct'; they disseminate standardized modes of regulation that ensure various locales are rendered visible, portable and combinable within the accounting network.

Given this apparent level of prescription, it is no surprise that commentators such as Miller and Rose (1990) and Rose and Miller (1992) have harnessed Latour's (1987) notion of 'action at a distance' to Foucaultian studies of governmentality. For instance, Nikolas Rose (1991: 675) believes accounting procedures and other techniques of numerical calculation have become 'integral to the problematisations that shape what is to be governed, to the programmes that seek to give effect to government, and to the unrelenting evaluation of the performance of government that characterises modern political culture' (1991: 675).[1] In Rose's view, numbers have become integral to modern government because they promise a privileged vantage point from which to view the domain to be governed. The collection of statistics – with its technologies for classifying and enumerating – allows civil domains to be rendered visible, calculable and, therefore, governable. Statistics are central to liberal government because they ensure that notionally 'free' subjects regulate their actions in ways delineated by the numerical techniques.

The governmentality perspective as set out by Rose and others implies that prescriptive mechanisms are ubiquitous and that space is strongly prescribed by powerful (governmental) networks. But as John Allen points out:

> The persuasiveness of this view depends not only on how much one believes in the actual immutability of what circulates between different settings at different times (is the meaning of such entities really so little dependent on context?), but also on the assumption that it is possible to replicate through translation strategies (in Latour's laboratory-like fashion?) the kind of schemas beloved of those in control. (2003: 133)

At the very least we should acknowledge that such replication is a precarious exercise, always subject to potential breakdown as networks struggle to 'normalize' behaviours in a range of settings. As Allen (2003: 137) puts it: 'an element of discretion and an independent use of power remain a constant possibility'.

This possibility is considered in work by Susan Leigh Star (1995). She insists that formalized arrangements do not arise simply from the imposition of prescriptive sets of relations by powerful network builders; instead, she argues, the effort that goes into making formalisms effective is usually invested in a series of 'trade-offs' between *generality* and *local uniqueness*, that is, network builders and enrolled entities must reach agreements or compromises if the network is be made stable through space.

Bowker and Star build on this insight to suggest that, from the point of view of the network builder, the creation of a perfect network often entails that some potential for discretionary action remains in the enrolled locale: the network 'ideally preserves common-sense control, enhances comparability in the right places, and makes visible what is wrongly invisible, leaving justly discretionary judgement'. They also say the network should retain 'intimacy (in its detailed knowledge of the nuances of practice), immutability-standardization, and manage[ability]' (2000: 232).

We can see, then, that an alliance between network imperative and local discretion is often the best means of achieving network extension, even within formal network arrangements such as those associated with accounting and other numerical procedures. However, Bowker and Star emphasize that the successful construction of such alliances is far from easy to achieve. Of the perfect network alignment they say:

> Such a perfect scheme [...] does not exist [...] Maximizing visibility and high levels of control threaten intimacy; comparability and visibility pull against the manageability of the system; comparability and control work against standardization. For a [network] to be standardized, it needs to be comparable across sites and leave a margin of control for its users; however, both requirements are difficult to fulfil simultaneously [...] The combination of these two thus requires compromise. (2000: 232–3).

Bowker and Star (2000: 292) therefore focus upon what they call (following Suchman and Trigg, 1993) the 'artful integration' of standardizing network and local action.

BOX 4.1

The relationship between network and local space can be orchestrated in three main ways:

- The network simply enrols the local and fully integrates it into existing relationships, leaving no room for localized discretion. Questions inevitably arise about the long-term effectiveness of such dominance, although, as Foucault has shown, such relations undoubtedly exist.
- The network becomes immersed in the local and loses its shape and reach. Again, there seems little long-term future for this network as the localities can too easily go their own way.
- A genuine interaction between network and local context takes place so that both are changed. This seems most likely to be the norm. However, it tells us little about the terms of enrolment as many variations on this arrangement might come into being.

In general terms, the varied interactions between networks and space illustrate the 'agonistic' relationship that tends to pertain between network space and geographical space. In particular, elements of the locale are selectively encompassed within the network and elements in the network are selectively grounded in the locale. Thus, Bowker and Star (2000: 307) conclude: 'things and people are always multiple, although that multiplicity may be obfuscated by standardised inscriptions'.[2]

We can therefore see that one form of multiplicity stems from the varied outcomes that ensue as networks move into new spatial locations and as agreements are reached between network builders and enrolled entities. The means by which the movement is made will be some combination of the enrolment procedures used by the network (often determined by the network centre) and the specific exigencies of the locale to be enrolled (the ways in which procedures of enrolment are tailored to distinctive features of the locality). As Law puts it:

> we need to hold onto the idea that the agent – the 'actor' of the actor-network' – is an agent, a centre, a planner, a designer, only to the extent that matters are also decentred, unplanned, underdesigned. To put it more strongly, we need to recognise that to make a centre is to be made by a noncentre, a distribution of the conditions of possibility that is both present and not present. (2002: 136)

However, there is another source of multiplicity that emerges from the interaction between networks and space. This refers to the *combined* effects of *multiple networks*. As differing networks come together in specific spatial locations so they generate outcomes that either reinforce singularity or give rise to multiple spatial identities. In order to generate singularities, the networks have to somehow 'punctualize' spatial identity – that is, they need to cut 'a specific figure in the here and now' (Munro, 2004: 294). In Munro's view, this process of 'punctualization' has two main aspects: first, there a 'positioning effect', so that the spatial entity is stabilized in the networks in a way that highlights or foregrounds particular features and characteristics. Second, there is a 'timing effect', in which the foregrounded identity displaces other potential identities in the 'here and now'. If these two effects are successfully brought together, a singular identity is generated. And if this identity can be stabilized within the networks, it may become enduring, perhaps even dominant. As often as not, however, the interaction between networks gives rise to a multiplicity, as Mol and Law emphasize when they say 'various "orderings" of similar objects, topics, fields, do not always reinforce the same simplicities or impose the same silences. Instead they may work – and relate – in different ways' (2002: 7). Thus, the two aspects of 'punctualization' mentioned by Munro may fail to achieve coherence: the networks cannot 'position' the space for any length of time; thus, alternative positionings co-exist and compete.

These observations indicate that we should not assume that spaces hold only singular identities – for instance, 'central', 'marginal', 'dominant', 'resistant' – rather, they can combine multiple processes, relations, identities, material arrangements and so forth (Hetherington, 1997). Thus, we should aim to develop a relatively sophisticated array of spatial typologies and we should consider how these interact with differing network arrangements. This suggests a need to investigate 'ways of describing the world while keeping it open, ways of paying tribute to complexities, which are always there, somewhere, elsewhere, untamed' (Mol and Law, 2002: 16).

Tracing a network topology

Of all the actor-network theorists, it is John Law who has put most effort into thinking through the spatial consequences of network relations. Like Latour, Law believes that actor-network theory poses a profound challenge to common-sense, taken-for-granted notions of space (Law, 1999). In Law's view, actor-network theory aims to establish a network ontology in which spatial formations are seen as *constituted by* heterogeneous sets of relations. In outlining this network ontology, Law begins by attacking the most common spatial type – that is, discrete, bounded Euclidian space. As Law and Hetherington put it:

in six-hundred years of surveying, cartography, nation-building and GIS, the idea that there is (a single) geographical space has been naturalised by Euro-Americans. This means that it has been very difficult to imagine space as anything other than some kind of neutral container, a medium, within which places [...] may be located. And this in turn means that any attempt to challenge this picture is very hard work and runs against the grain of common sense. (1998: 9)

Relationalism, Law argues, runs counter to the notion that there is single, bounded space in which things simply happen and actor–network theory has been developed in order to show the complexity of spatial relationships and the multiplicity of spatial types. In Law's view, actor–network theory shows that space is not itself a *container* but is *contained* (in networks). Thus, space is no longer singular in character but consists of varied space-times, all operating in differential spatial configurations.

Law suggests that we should abandon *topographical* notions of space – in which the space of absolute and fixed coordinates is necessarily dominant – in favour of *topological* conceptions.[3] He sees topology as concerned, in the main, with the way spatial objects are both constituted and displaced by networks; as Mol and Law (1994: 643) say: 'topology doesn't localise objects in terms of a given set of coordinates. Instead, it articulates different rules for localising in a variety of coordinates'. Mol and Law argue that in topological space we can discern differing spatial types to those found in topographical configurations. There are of course regional spaces in which 'space is exclusive. Neat divisions, no overlap. Here or there, each place [...] localised on one side of the boundary' (1994: 647). However, actor–network theory tells us that even in these regional or exclusive spaces, spatial relations are *performed* by networks: 'Space is made. It is a creation. It is a material outcome. Like objects or obligatory points of passage, it is an *effect*' (Law and Hetherington, 1998: 8). Thus, as well as regional spaces, we should expect to find network spaces. These differ sharply from Euclidian spaces because 'in a network, elements retain their spatial integrity by virtue of their position in a set of links or relations. Object integrity, then, is not about a volume within a larger Euclidean volume. It is rather about holding patterns of links stable' (Law, 1999: 6).[4]

The recognition that networks generate their own specific space-time configurations leads inevitably to a network topology, which is seen as an undulating landscape in which the linkages established in networks draw some locations together while at the same time pushing others further apart. This network topology can be discerned in the following comment by Latour.

> In a network certain very distant points can find themselves connected, whilst others that were neighbours are far removed from one another. Though each actor is local, it can move from place to place, at least as long as it is able to negotiate equivalences that make one place the same as another. A network can thus be 'quite general' without ever having to pass through a 'universal'. However rarefied and convoluted a network may be it nevertheless remains local and circumscribed, thin and fragile, interspersed by space. (1987: 170–1)

The landscape is 'folded', 'pleated' and 'ruptured' by the spacing and timing activities of networks that run through, around or underneath it. In this network space 'proximity isn't metric'; rather, it has to do with 'the network elements and the way they hang together. Places with a similar set of elements and similar relations between them are close to one another and those with different elements or relations are far apart' (Mol and Law, 1994: 649). Thus, distance is 'a function of the relations between the elements' (1994: 643).

As we have seen, a network topology inevitably disturbs our received ideas about (Euclidean) space. However, having asserted the importance of this new perspective, actor-network theorists now propose that topological complexity cannot be adequately comprehended simply through the prism of the network. In short, it is argued that network space does not exhaust the range of spatial possibilities that might emerge. In part, this concern over the status of network space stems from a concern over the status of the term 'network' itself. In an influential commentary on actor-network theory, Lee and Brown (1994) argue that the network concept has a tendency to imperialistically colonize all domains so that ultimately nothing can stand outside actor-networks (or, for that matter, actor-network *theory*). Lee and Brown suggest that the approach has moved in this all-encompassing direction because it weaves together a Nietzschean concern for 'the will to power' (through the building of networks in the style of Pasteur) with a liberal democratic notion of 'enfranchisement', that is, extending agency to all things (Latour's immutable mobiles). As Lee and Brown note, there is no space outside the network; simply endlessly ramifying network relations which appear to leave no hope of escape to a zone beyond translation and enrolment.

In response to this and similar criticisms – notably from feminist theorists such as Susan Leigh Star (1991) and Donna Haraway (1997), who complain that actor-network theory has tended to focus its attention on the network builders (such as Pasteur) rather than on those systematically excluded from network relations – Mol and Law outline another spatial type which they call 'fluid space'. In their view, fluid space stands in stark contrast to network space. Where, in a network, the relations between actors, entities and objects are clearly defined, in a fluid space there is no such clear definition either in the relations or in the shape of the enrolled elements: 'in a network things that go together depend on one another. If you take one away, the consequences are likely to be disastrous. In a fluid space it isn't like that because there is no "obligatory point of passage"; no place past which everything else has to file; no Panopticon; no centre of translation' (Moll and Law, 1994: 661). Thus, 'in a fluid space it's not possible to determine identities nice and neatly, once and for all'. Instead, all we find in this space are 'viscous combinations' in which 'elements inform each other' in ways that 'continuously alter' (1994: 660). Yet, despite the viscosity, Law and Mol (2000: 6) emphasize: 'fluid spatiality [...] rather than representing breakdown or failure, may also help to strengthen objects'. Fluid relations, although quite distinct from regional and network forms, may therefore represent enduring features of the complex topologies that now compose the spatial realm.

BOX 4.2

Mol and Law (1994) introduce three main spatial types:

- 'Euclidean' or 'topographical' space. This refers to spaces of fixed coordinates, with lines that run across surfaces (rather in the style of maps, with their contours and two-dimensional spatial representations). Mol and Law's criticisms of this spatial type echo Doel's (1999) criticisms of what he calls 'pointillism' (discussed in Chapter 1) – that is, a concern for surface and lines between points leads to only a superficial understanding of spatial relations.
- 'Network' space. This is the space of actor-network theory, especially in discussions of well-orchestrated, tightly knit networks, as in the case of Pasteur (discussed in Chapter 3). Network space is composed from the heterogeneous relations normally assembled within actor-networks. Mol and Law's concern about network space is that it can focus too much attention on tightly structured modes of ordering space.
- 'Fluid' space. This is a new focus for theorists working in the actor-network genre and refers to spatial relations that are constantly 'becoming', constantly shifting, constantly moving. This spatial type fits well with the notion of spaces of multiplicity which is so central to post-structuralist geography.

Mol and Law introduce this idea of 'fluid space' to counter the hegemonic tendencies of network space, the belief that network builders will ultimately triumph in imposing singular identities on multiple participants. Rather, they want to make room for difference and diversity. As Law summarizes it, multiplicity means

> more than one and less than many. Fractional natures. Fractional and enacted bodies. The webs in which they are enacted are partially other. Other, as it were in general, but also to each other. And their relations are uncertain. Perhaps sometimes, they fit together neatly. Perhaps they contradict one another. Perhaps they pass each other by without touching, like ships in the night. Perhaps they are included in one another. Perhaps they are added together to produce new natures. Perhaps they are deliberately kept apart because any encounter would be a collision. Or perhaps their relations are a mix of these: complementary, contradictory and mutually inclusive. At any event, in this way of thinking natureculturetechnics are complex in the sense that they are multiply enacted in multiple practices and cannot be known anywhere in particular. (2004: 6)

Such formulations indicate that the actor–network theorists are moving away from a concern simply for the *centring* practices that permit actors to become powerful in networks (in response to the criticism that it thereby focuses too much upon the *already* powerful). They are now attempting to consider both inclusions and exclusions as networks are formed, that is, 'the oscillation, absence/presence, uncertainty and [...] necessary Otherness that comes with the project of centering' (Law, 2002: 136–7). As well as this emphasis on network exclusions, we also see an increasing interest in the spatial effects of contextual relations. It is now recognized that subjects and objects are drawn selectively into and out of discrete networks, sometimes simultaneously, sometimes concurrently. This leads to greater spatial complexity as spaces emerge not from one (centred) network space but from *multiple* spatial forms, all working within *contexts of multiplicity*.

Geo-philosophies of relationalism

Complexity, multiplicity and topology are now increasingly combined in actor-network theory in order to escape the rather restricted modes of spatial ordering that emerge from studies of Panopticism and other highly prescriptive network forms. The analysis presented in the previous sections suggests that there are varied network types and varied relations between these types and spatial locations. So while we can retain the view that space emerges from within networks, we can now suggest that it does so through some complex interactions between the network and those entities and spaces that lie 'outside' it. In short, the network and its (spatial) environment mutually compose one another, often in varied and unexpected ways.

Thus, network spaces might be placed on a continuum. At one end, we have 'singular' spaces in which formal and standardized sets of relations succeed in marking out clearly demarcated zones where entities and actors are both stabilized and normalized in topographical fashion. These formalized and standardized relations can be generated by singular networks or by the co-ordinated actions of multiple networks 'meeting in' space. At the other end, we have highly fluid spaces in which flux and variation are the norm as actors and entities struggle to impose coherence onto multiple relations. Again, fluidity can be the property of a single network or can arise from the combined effect of multiple network interactions. As we move along the continuum, differing trade-offs between singularity and multiplicity, topography and topology might be observed.

In this section, we delve further into this characterization of network topology. We explore in a little more detail the complex network forms that emerge as we move into topological space. In so doing, we also move beyond Foucaultian/ actor-network theory in order to explore the spaces of Gilles Deleuze and Michel Serres. Both these philosophers have exercised some considerable influence over post-structuralist geography (as well as over actor-network theory) and it is worth briefly considering the nature of their influence, especially as it relates to notions of topology and space.[5] However, as we shall see, a brief consideration of spatial topology in the work of Deleuze and Serres rather unexpectedly brings us back to more traditional geographical notions of territorial space. That is, Deleuze and Serres tend to adopt general abstract perspectives on the spatial realm, perspectives that are apparently developed over and above perspectives derived from within specific network formations.

Gilles Deleuze, while a key philosophical influence on actor-network theory, is also a key theorist of the relations between complexity, multiplicity and space. The connections between these three terms are well explained in Manual Delanda's (2002) exposition on Deleuze's 'philosophy of becoming'. In Delanda's view, multiplicity emerges naturally from Deleuze's fluid and dynamic ontology. Instead of assuming fixed and invariant essences, Deleuze considers 'being' in the world to be based on movement and emergence. As Delanda (2002: 3) puts it, 'Deleuze is not a realist about essences, or any other *transcendent* entity, so in his philosophy something else is needed to explain what gives objects their identity and what preserves their identity through time. Briefly, this something else is *dynamical processes*' (emphasis in original). Deleuze's interest in processes of 'becoming' leads, in turn, to an interest in topology. Delanda describes Deleuze's topological perspective in the following way:

> [topology] may be roughly said to concern the properties of geometric figures which remain invariant under bending, stretching or deforming transformations which do not create new points or fuse existing ones. (More exactly, topology involves transformations [...] which convert nearby points into nearby points and which can be reversed or be continuously undone.) Under these transformations many figures which are completely distinct in Euclidean geometry [...] become one and the same figure, since they can be deformed into one another. (2002: 25–6)

In this topological field, the issue that preoccupies Deleuze is 'how to conceive of a form of identity or unity which is not identical to itself' (Patton, 2000: 29). That is, Deleuze wishes to move beyond simple repetition and resemblance in order to study difference and divergence, as he makes clear in the following geographically-inspired quotation.

> Maps [...] are superimposed in such a way that each map finds itself modified in the following map, rather than finding its origin in the preceding one: from one map to the next, it is not a matter of searching for an origin, but of evolutionary displacements. Every map is a redistribution of impasses and breakthroughs, of thresholds and enclosures, which necessarily go from bottom to top. There is not only a renewal of directions, but also a difference in nature: the unconscious no longer deals with persons and objects but with *trajectory* and *becoming*: it is no longer an unconscious of commemoration but one of mobilisation, an unconscious whose objects take flight rather than remaining buried in the ground. (Deleuze, quoted in Crang and Thrift, 2000: 21)

These 'evolutionary displacements' stem from the play of differences as movements are made from one stage (of mapping) to the next. Thus, in Deleuze's philosophical world,

> becoming necessarily entails deformation, reformation, performation, and transformation, which involve gaps and gasps, stutters and cuts, misfires and stoppages, unintended

outcomes, unprecedented transferences, and jagged edges. These breaks are not simply ungoverned transversal communications within and between assemblages that bring novel forces into play and so also new formations. They are also a function of the way events occur, which is not rule governed, or where the rule does not apply. So, Deleuze stresses connectivity of systems in opposition to what he regards as an illusory autonomy promoted by some writers. (Thrift and Dewsbury, 2000: 418)

In this context, Deleuze insists on the importance of multiplicity as a means of accounting for both invariance and transformation. Multiplicities emerge as singular locations and come together in 'recurrent sequences' (Delanda, 2002: 16).[6] These sequences involve 'active transformations' in a process which 'converts one of the entities into the other' (2002: 18). As Deleuze puts it: 'the actualisation that stabilises and stratifies [...] is an integration: an operation which consists of tracing "a line of general force", linking, aligning and homogenising particular series and making them converge. There is [...] a multiplicity of local and partial integrations, each one entertaining an affinity with certain relations or particular points' (1988: 75).

We can hear in these comments clear echoes of actor-network theory – for instance, in the emphasis upon emergent properties, the attention to sequences ('networks'), transformations ('translation'), and the engagement with an open-ended topological complexity ('generated by networks'). Moreover, both Deleuze and the actor-network theorists see relation and space as co-emergent: as Ansell-Pearson (2002: 24) puts it, space cannot be taken to be an *a priori* reality but must be seen as an 'emergent and exigent feature of social action'. This observation bears upon a key question raised by John Law: 'do networks subsist in and of themselves? Are they, as the actor-network theorists have tended to assume, spatially autonomous?' (2000: 8). Law answers these questions in Deleuzian fashion by claiming that networks cannot be seen as somehow separate from spatial relations, for the relationship between network and space is always reciprocal; the two compose one another in mutually reinforcing ways. Thus, as Doel explains:

> It would be better to approach space as a verb rather than a noun. *To space* – that's all. Spacing is an action, an event, and a way of being. There is neither space 'behind' something, functioning as a backcloth, ground or continuous and unlimited expanse (absolute space), nor space 'between' something, as either a passive filling or an active medium of (ex)change (relative, relational, diacritical, and dialectical spaces). There is just spacing (differentials). The 'points' – as things, events, terms, positions, relata, etcetera – that are supposedly played out 'upon' and alongside space are illusory. Space is immanent. It has only itself. (2000: 125)

Like actor-network theory, Deleuzian theory discusses how differing sets of relations give rise to differing spaces. For instance, it distinguishes 'linear' relations, which present simple distributions of points, from 'nonlinear' relations, which display multiple connections.[7] Alternatively, it distinguishes between 'assemblages of desire that are fixed or delimited in particular ways, shut off from all but certain specified relations to the outside' and 'more fluid and open-ended assemblages in which new connections and new forms of relation to the outside are always possible, even at the risk of transforming the assemblage into some other type of body' (Patton, 2000: 43). We see here something of a reworking of centred and decentred actor-networks, with the former generating spatial forms of fixed coordinates while the latter give rise to fluid, viscous combinations.

BOX 4.3

For Deleuze, space is:

- In the process of 'becoming'. It results from dynamical processes and it is always on some kind of emergent trajectory.
- Subject to transformation. Its reproduction within processes of becoming is based not on simple replication but on alteration and innovation. Entities are folded into one another as new relations come into being. Thus, space takes on new shapes and new identities; it is always emergent.
- Multiple in nature, it is generated in 'recurrent sequences'. These sequences are generated within spatial trajectories that can either create further multiplicities or can result in unities of various kinds.
- Moreover, differing trajectories or lines of force hold differing consequences for territories; they can result in deterritorialization or reterritorialization. Thus, lines of force can work to unify territorial spaces (perhaps using processes of governmentality) or can work can to disrupt territorial coherence thereby revealing multiplicities of various kinds.

In their different ways both actor-network theory and Deleuzian theory bring us to spaces of multiplicity: 'On the one hand, multiplicities that are extensive, divisible and molar; unifiable, totalisable, organisable [...] and on the other hand, libidinal, unconscious molecular, intensive multiplicities composed of particles that do not divide without changing in nature, and distances that do not vary without entering another multiplicity' (Deleuze and Guattari, 1987: 33). Multiplicities conceive spatial forms through their generative capacities, and these depend on the emergent properties (or 'affects') that come into being as relations are formed between entities of various kinds. They compose 'a nested set of spaces, with the cascade acting to unfold spaces which are embedded into one another [...] what matters about each space is its way of being affected (or not affected) by specific operations themselves characterised by their capacity to affect (to translate, rotate, project, bend, fold, stretch)' (Patton, 2000: 69).

Deleuze (with Guattari) emphasizes that these differing assemblages hold differing relations to territory. Assemblages of fluid and viscous forms tend towards *deterritorialization* – that is, 'lines of flight along which the assemblage breaks down or becomes transformed into something else' (Patton, 2000: 54). This movement generates, however, a counter movement, a *reterritorialization*, an effort to resituate assemblages in a defined space of fixed coordinates (a map, an administrative zone and so forth). The first of these movements highlights the new spatial forms that can emerge either through network modification or the coming together of two networks to generate a new form of 'becoming'. The second refers to those normalizing and governmental networks that Foucault sees working in liberal society to 'fix' (that is, to 'territorialize') the 'conduct of conduct'. As in Foucaultian/actor-network theory, we can discern a 'two-way' movement here as differing relations work to confine or produce spatial multiplicities.

An illustration of how differing assemblages underpin specific territories is supplied by Bonta and Protevi (2004) in a Deleuzian study of the Olancho region in Honduras. The authors suggest that multiple spaces are evident in this territory:

> There was no a priori 'Olancho Space' that then got broken down into smaller side-by-side, nested-hexagon spaces; you didn't walk from one space to another as much as move through varied degrees of becoming across the landscape. Coffee farms were being taken over by cattle; beans were taking over forest; forests were taking over ranches; Hurricane Mitch had stripped away cattle pastures, beanfields, coffee, and forest alike. A 'space', then, became the room filled by the workings of a complex system at the extensive, intensive, and virtual registers. If one were 'plugged into the cattle-ranching assemblage, one was to a large extent predetermined and at the very least codetermined by a complex system quite different than that of one's neighbour, who was plugged into the complex system of coffee farming – or of peasant farming, conservation, development, logging, and so forth. One was an 'actor' in that complex system's space – an enactor of its space – wherever one went. Inasmuch as one (and one's cattle or coffee bushes or beans) 'made space', one territorialized one's assemblage somewhere in the landscape. One carved out a territory to provide room 'demanded' by one's assemblage. (2004: 172)

As in actor-network theory, space is here relationally constituted by assemblages that pull certain places into proximity while pushing others into the distance (with distance conceptualized as relations between the aligned elements). And again, as actor-network theory emphasizes, differing spatial forms inhabit the same territory:

> Each space is qualitatively different – they are not variants on one of them, but come about through vastly divergent processes. Each has a geohistory that must be engaged on its own terms. Each is territorialized in the landscape by means of human and non-human 'agents' guided by a certain set of instructions, tendencies, trajectories [...] Each complex system deterritorialises forces of the earth and the cosmos and puts them to work in a different way, stratifying them in different sequences, drawing from elements common to them – on the physico-chemical, geological, biological, and human strata – but for different purposes. (Bonta and Protevi, 2004: 173–4)

The language is Deleuzean, but we see here clear affinities with Foucaultian/actor-network theory's heterogeneous networks, weaving materials of differing kinds into new spatial formations. These formations are plural and multiple, comprising striated and smooth spaces, prescriptive and fluid relations, territorialized and deterritorialized assemblages. However, Bonta and Protevi's characterization of Olancho also takes us a little closer to more standard geographical concerns, notably the requirement to make some general comments about the nature of territories situated within given spatial contexts. Although they highlight the differential nature of the assemblages in Olancho, Bonta and Protevi also give us a kind of territorial overview of this place, one that highlights how multiple relations co-exist within its boundaries. They provide us with what Michel Serres calls a landscape of 'nearness and rifts' (Serres and Latour, 1995: 60), a scene in which entities are pulled closely together within specific assemblages while other entities, which are proximately situated in Euclidean space, are pushed away into alternative assemblages. In this landscape, our attention is drawn to the 'stopping points, ruptures, deep wells [...] rendings, gaps' that create proximities and rifts between assemblages, networks and entities (Serres and Latour, 1995: 57). Serres suggests that we might view such landscapes in the following way.

> If you take a handkerchief and spread it out in order to iron it, you can see certain fixed distances and proximities. If you sketch a circle in one area, you can mark out nearby points and measure far-off distances. Then take the same handkerchief and crumple it, by putting it in your pocket. Two distant points suddenly are close, even superimposed. If further, you tear it in certain places, two points that were close can become very distant. This science of nearness and rifts is called topology, while the science of stable and well-defined distances is called metrical geometry. (Serres and Latour, 1995: 60)

Again, the notion of topology is being used to allude to the stratifications that 'fold' and 'pleat' space (the mountainous range of valleys and peaks). Space, like time, is folded into complex geometries as networks draw points to the surface and push others underground.

BOX 4.4

For Serres, space is also:

- Multiple in nature, so that various processes run through and around space, giving rise to undulating landscapes of spatial relations.
- Made up of entities that are bound into relations that can bring the 'far near' and make the 'near far'. Thus, space is wholly relational rather than absolute; any given space is thus a complex arrangement of differentially articulated relationships.

Continued

- Brought into some kind of unity by the 'relations between relations'. The messengers and communicators that map out the spaces 'in-between' relations can create new proximities and can generate some amount of coherence and stability, thereby giving rise to discernible territorial shape.
- Turbulent and chaotic in nature so that networks can only create temporary permanences (or islands of stability) with the danger that they will dissolve once again into the disordered flow of space-time. In order to prevent such dissolution networks may use either topographical or topological modes of ordering.

While we find echoes of Deleuze's smooth and striated spaces and actor-network theory's spaces of singularity and multiplicity in this formulation, we also discern the emergence of a broad perspective that attempts to take in *all* the assemblages and networks, that is, those that aim at *de*territorialization and those that aim at *re*territorialization. Serres, like Foucault, wishes to understand how the connections between sites and structures work to generate 'unities' (Gutting, 2001: 233). In order to discern these unities, he aims to place himself in a position where he can freely criss-cross the intermediate zones, drawing out from the networks what best illuminates the entire landscape. This desire to chart the 'cross-over', to be situated *between* assemblages, networks, lines of flight, stems from Serres's efforts to reformulate some long-standing and accepted conceptual divisions, such as those between self and society, subject and object, science and literature, the social and the natural (see Serres and Latour, 1995). According to Brown (2002: 1–2), Serres 'proceeds from the notion that disciplinary and conceptual divisions, although complex and provisional, may be analysed by exploring potential channels or "passages" that run between them'. In making these 'border crossings' Serres aims to show how the world is dis/ordered – that is, he aims to tell us something tangible and clear about things that are missed by mainstream disciplinary knowledges. He aims at a 'method of rapid movement, and congruent "comparativism" […] the method of the space between, of conjunction of bringing into proximity' (Bingham and Thrift, 2000: 285).

As indicated above, we still encounter here a topology based on multiple space-time relations. However, Serres is concerned to philosophically (re)construct landscapes in which assemblages and networks 'communicate' with one another. Communication 'traverses those spaces […] that are much less clear and transparent than one would have believed' (Serres and Latour, 1995: 75). He therefore explains a need to 'describe the space situated between things that are already marked out' (Serres and Latour, (1995: 64). Spaces in-between the assemblages, the orderings, the networks are, Serres believes, 'more complicated than one thinks. That is why I have compared them to the North West Passage […] with shores, islands, and fractal ice floes' (Serres and Latour, 1995: 70). However, while he focuses on the in-between, Serres is also interested in discerning some kind of connectivity across the divisions. He does not want to

be confined to existing or consolidated relations; he wishes to explore an area
that he believes 'is strangely devoid of explorers' (Serres and Latour, 1995: 70)
in order to discern the 'network of multiple bonds', the 'lattice of relations'
(Serres, 1995: 111) that ultimately tie local domains into a 'global landscape'
(Serres and Latour, 1995: 118). This space 'in-between' is, for Serres, a potentially chaotic turbulent place; in fact, chaos and turbulence appear to be *primary*,
while spaces of organisation and stability are *secondary* (Conley, 1997: 62).
Network builders must therefore struggle *against* disorder and disarray. We might,
then, see stabilized networks as 'islands' set within a broader context of commotion and flux.

In these observations Serres directs our attention to the way that networks
define themselves against their contexts of emergence. As processes of definition unfold, networks can become either relatively closed – thereby establishing sharp boundaries between their own internal relations and contextual
relations – or they become relatively open – thereby ensuring fluid interactions
between internally and externally constituted relations. Networks will adopt
whichever strategy ensures stability. Space will thus be constituted in very different ways depending on the network type. If networks are 'closed' then space
is likely to be constrained and bounded, confined within Panoptical sets of
relations. If the network is 'open' then space is likely to be emergent and fluid,
channelled along and through multiple lines of flight. In other words, the interaction between network and context works to generate either singularities or
multiplicities. As Doel, using Deluzian terminology, puts it: 'A fold is always at
least twofold. Sometimes it functions as a line of rigid or supple segmentation,
which effectively partitions and territorialises the plane of immanence into a
plane of organisation. Sometimes it acts as a line of flight, which unfolds and
deterritorialises the plane of organisation' (1999: 165).

At certain times and in certain places, multiplicities are 'folded' *into* singularities, while at other times and in other places singularities are 'unfolded' into
multiplicities. For instance, in Foucault's Panoptic prison the process of folding
in is seemingly so well executed that ordered and predictable outcomes are
routinized in the actions of potentially unruly prisoners. A tight lattice of relations is consolidated and this keeps both internal dissent or external interference at bay. However, the lattice can also unravel so that the repressed or
contained multiplicities begin to unfold outwards within a myriad of lines or
networks. At this point the division between the prison and its contextual environment begins to dissolve and topological relations displace topographical
relations.

To summarize this section, we can say that while the theories put forward
by Deleuze and Serres vary in the language they use to describe assemblages,
networks, lines of flight and so forth, they both point us towards space as an
outcome of relations. These relations are sometimes made in tightly controlled
ways, while at other times they are fluid and viscous. As they run through space
they weave patterns in the landscape, drawing some places together, pushing

others apart. They create both proximities and distances, topographies and topologies. Deleuze and Serres provide rich descriptions of such spatial variations. They also recognize that the areas 'in-between' are a prominent aspect of post-structuralist geography. These 'in-between' spaces, which are turbulent, unstable and viscous, must be carefully navigated by the networks. Two main navigational strategies present themselves: first, draw sharp distinctions between network and context so the turbulence is kept at bay; second, constitute the network in such a fashion that it somehow 'internalizes' flux and instability. In practice we might suggest that some combination of the two strategies represents the norm.

In order to assess the differing spatial effects of network strategies Deleuze and Serres encourage us to seek out *vantage points* from which we might observe *all* the networks, assemblages, lines of flight. In this regard, they return us to *geography*, to the analysis of spatial formations that are simultaneously ordered and disordered. They emphasize that processes of (dis)ordering emerge from the activities of multiple networks running through and around specific spatial arenas. They therefore suggest we assess 'a double articulation of incompossibilities: the smooth and striated; territorialisation and deterritorialisation; stabilisation and destabilisation; constancy and consistency' (Doel, 2000: 124). In these various combinations we find the proper disposition of relational space.

Conclusion

Over the last three chapters, we have moved further and further into relational space. We began with Foucault's explorations of discursive space, the way spatial arrangements 'mirror' discursive formations. We saw that as Foucault's work progressed, he paid more and more attention to the materiality of discourse, to the way discourse becomes encoded or embedded within the material arrangements that comprise prisons, asylums and other such institutions. He also reflected on practice, in particular on the strategies of normalization that seem to accompany the assembling of materials in the shape of institutional sites. He explained that these materials are constituted in ways that bear down upon individual behaviour in order to both generate knowledge about this behaviour and to prescribe the actions that individuals can take (these two aspects come together particularly clearly in the Panoptic prison).

Space is here made within sets of heterogeneous materials, all assembled in line with certain discursive priorities. The same theme is evident in actor-network theory, which also explores how space is made in heterogeneous ways. Again, the emphasis is on relations assembled in line with particular strategies of normalization, this time devised in clearly defined centres of calculation (such as laboratories). Actor-network theory extends the range of discursive rationalities in play by attending to the work of scientists as they disseminate their knowledges and their artefacts. The theory also problematizes the interaction between discourse and alignments of heterogneous materials, for it claims that these materials cannot just be seen as 'effects' of the discourse; rather, they play a real and active role in the materialization, and thus the ultimate shape of the discourse itself. In short, discourse and material enter a reciprocal relation.

In many ways, the discussion of Foucault's work in Chapter 2 and the discussion of actor-network theory in Chapter 3 cover similar ground: they both show how actors become powerful (the guards in the Panoptic prison, Pasteur in his laboratory) and they outline the importance of spatial arrangements in generating this power. Actor-network theory in particular provides a rich description of the means whereby power relations facilitate movement from one place to another. It therefore allows us to see how spatial relations become intrinsic to power relations (that is the true significance of 'action at a distance'). Actor-network theory also makes the strong claim that relations always run at 'ground level': we never need to shift scale from the 'micro' to the 'macro' or from the 'local' to 'global'; rather we need to attend to network length. Thus, the methodological priority is to follow the networks wherever they might lead, illustrating how they 'space' and 'time' as they go (Bingham and Thrift, 2000).

Yet, the accounts provided in Chapters 2 and 3 have a tendency to focus on normalizing and prescriptive sets of relations. As we have seen, this focus can lead to unwarranted assumptions about the ease with which networks enrol both actors and localities into their modes of functioning. In so doing, the theory downplays one of Foucault's key insights – that any extension of power rela-tions will inevitably meet resistance. The interplay between extension and resis-tance takes the form of an 'agonism' as the two forces struggle for supremacy. A number of those working in the Foucaultian/actor-network theory tradition have now begun to investigate the spatial effects of this 'agonism'. In particu-lar, they have begun to show how networks and localities genuinely 'interact' so that some some modification of both partners – the network and the locale – takes place. This can be seen as a kind of 'trade-off' between network space and local space. The upshot is that network relations should be seen as a mixture of local specificities and network regularities. Networks, it might be argued, come in a variety of shapes and sizes, as do network spaces. We therefore need to look in detail at how networks operate, how they move from place to place, and the types of relationships that are established between network form and spatial location.

It has been proposed that while the in-depth analysis of networks is a good means of understanding the relational construction of space, we also need to stand outside the networks in order to see the broad spatial terrain. This returns us to notions of geography as a kind of 'imperialistic' science, one that attempts to understand the spatial unities that encompass multiple sets of (networked) rela-tions. The reassertion of this more traditional geographical perspective suggests a need to combine both topographical and topological conceptions of space: that is, we should accept that space is generated from within sets of (networked) rela-tions but we should also recognize that these relations must be situated within broader contexts of movement and flux. We should therefore aim to hold the multiple and the singular together: we need to consider how varied relations run 'through' space, weaving their own space-time trajectories, and we also need to consider how these relations interact with their broader spatial environments. In this regard we see that some networks aim to establish sharp boundaries between the 'inside' and the 'outside' while others remain fluid and permeable, open to outside influence and accommodating of externally-inspired change. These differing network types can be seen as 'singular' or 'multiple', 'reterritorialized' or 'deterritorialized'.

We therefore arrive at considerable spatial complexity. In order to describe this complexity, a whole host of terms have been employed – singularity, multiplicity, fluidity, network, topology, assemblage, line of flight, territory, and so forth. Some of these terms are clearly more useful than others, but undoubtedly all retain an abstract quality, as though the effort to describe complex relational spaces defies more commonplace speech. In superficial terms, this observation seems credible: relationalism means overthrowing or amending more traditional ideas about the spatial realm, and more often than not any new way of seeing requires some new way of talking. And yet there is more to it than this, for it may be that the abstract quality of the terms owes something to the sheer difficulty of talking about space *in the abstract*. Andrew Sayer, for one, believes this to be the case: he says that, in the main, 'spatial theory can make only vague allusions to particular kinds of spatio-temporal organisation'. In his view, 'only more concrete analyses can hope to say more' (2004: 268). In concluding the theoretical section of the book, it must be acknowledged that the 'vague allusions' outlined above may be far too vague for some readers' tastes. However, in charting the development of post-structuralism through the theories of Foucault, Latour, Law, Deleuze and Serres, we have seen where this language has come from and what it endeavours to describe. It has been elaborated here in the belief that it does indeed help us to think about space, and the role that geography might play in understanding space, a little differently. However, the significance of these ideas will only be clear once they are invoked in particular geographical contexts. With this in mind, subsequent chapters move down from the abstract level to the case-study level in order to illustrate how the post-structuralist theories reviewed above might be applied in research practice.

SUMMARY

In this chapter, we have investigated the topological spaces that emerge from multiple sets of relations. We began by considering networks of prescription and networks of fluidity and flux. The first network type tends to work with a formalized structure and therefore seeks to construct space in formal and prescriptive ways; the second type is only loosely assembled and permits a more undulating, topological space to come into being. We then turned to examine the relationship between network and territory in more detail, drawing upon work by Deleuze and Serres. It was shown that while space is underpinned by relations this does not mean that space is only relational in nature; territorial integrity and unity can still emerge. It was therefore concluded that we need to combine notions of demarcated enclosed spaces with processes of emergence and becoming.

FURTHER READING

A useful introduction to Law's general perspective on relationalism can be found in Law and Mol (eds) (2002) *Complexities: Social Studies of Knowledge Practices*. Geof Bowker and Susan Leigh Star's (2000) book, *Sorting Things Out: Classification and Practice*, considers some of the spatial issues that arise as networks are extended. There are many secondary texts on Gilles Deleuze. One of the most useful and accessible is Paul Patton's (2000) book, *Deleuze and the Political*. Some idea of what Deleuzianism means for geography can be gained from Marcus Doel's (1999) *Poststructualist Geographies*, as well as Mark Bonta and John Protevi's (2004) *Deleuze and Geophilosophy: a Guide and Glossary*. Anyone wishing to investigate the work of Michel Serres should probably start with the stimulating discussion between Serres and Latour in their (1995) book, *Conversations on Science, Culture and Time*.

Notes

1. This is evident, for instance, in Ian Hacking's work on statistics. In the *Taming of Chance* (1990) Hacking traces the emergence of statistical techniques in the eighteenth century (the census, official statistics, taxation statistics and so on) and shows that statistics have become central to modern forms of governance, so central in fact, that good government becomes almost unthinkable without them.
2. Bowker and Star give this concern an ethical slant when they suggest that we hold 'firmly to a relational vision of people–things–technologies' in order to assess how networks and local arrangements can be *made* to interact in ways that reflect the aspirations of the multiple constituencies that reside at the interface of the two domains.
3. Law takes this notion of 'topology' from mathematics but it seems to be inherited mostly from the work of Michel Serres (see Serres and Latour, 1995, and the discussion below).
4. This discussion of objects here derives from the importance actor-network theorists attribute to technology in the making of space. As Callon and Law put it:

> The importance of technologies for folding together places, actors or actants is obvious. Technologies and material arrangements distribute actions and actors. The local is never local. A site is a place where something happens and actions unfold

because it mobilises distant actants that are both absent and present. The drawing in the school exercise book illustrates this strange ontology. The blacksmith is walking beside the ploughman, his hand resting affectionately on his shoulder, but his silhouette is surrounded by a blue halo, just like a guardian angel. (2004: 6)

5.   It is worth noting that during a visit to London in the mid-1990s, Serres is reported as describing Deleuze and himself as 'philosophical geographers' while he characterized Foucault and Derrida as mere 'historians of philosophy' (quoted in Critchley, 1996).

6.   Moreover, the recurrent sequences that generate multiplicities give rise not to some closed, final, essential product but to divergent forms in a potentially endless series. Thus, a multiplicity has 'no need whatsoever of unity in order to form a system' (Deleuze, quoted in Delanda, 2002: 13). As Rajchman (2001: 60) says, a multiplicity is 'folded many times over and in many ways such that there is no completely unfolded state, only further bifurcations'.

7.   Linear relations tend to fix entities into stable shapes, described by Delanda (2002: 13) as 'essences' that 'possess a defining unity [...] and, moreover, are taken to exist in a transcendent space which serves as a container for them or in which they are embedded'. Non-linear, complex relations differ to the extent that they do not have 'a supplementary dimension to that which transposes upon it. This alone makes it natural and immanent'.

Part 2   Cases

Introduction

In this empirical part of the book, we examine case studies which illustrate some of the theoretical issues raised in the first section. At the outset it should be made clear that these cases *do not* comprise simple applications of the theoretical frameworks emerging from the first part of the book. Rather they serve to show how *post-structural* spaces – for instance, spaces of singularity and multiplicity – are currently emerging within mainstream geographical research arenas.

The case-study chapters focus particular attention on the problematic interaction between complex, heterogeneous *processes* and coherent, stabilized *territories*. In so doing, they investigate efforts to 'order' space – that is, they show how various social actors seek to 'ground' complex processes in coherent and robust spatial arrangements. Thus, the spatial arrangements that come to dominate in any given instance can be seen as the outcome of struggles to impose stability in contexts of flux and fluidity. The chapters indicate that, more often than not, dynamic relations work to undermine formally constituted territories. Nevertheless, the struggle to 'order' space is necessarily on-going, as relations are continually harnessed to the process of building new spatial formations.

In the first case-study chapter, we examine the status of nature in a relational world. We investigate efforts in the post-war period to 'contain' nature in a strictly demarcated zone (in the English countryside) in order to protect it from dynamic and heterogeneous processes. We see that as nature came to be spatially bounded, transgressive processes began to work across the boundary line. These transgressive processes clearly threatened the protected (or zoned) nature but, paradoxically, they also worked to secure a more robust division between nature and society. The chapter shows that while nature should be seen in relational (notably, ecological) terms, efforts at 'containing' nature are still required if natural entities are to be sustained through time. In other words, while space may be made of relations we still need to ensure some degree of spatial permanence. Thus, ways must be found to align topographical and the topological spaces within post-structuralist accounts.

In the next case-study chapter, we look a little more closely at planning in order to see whether this form of governmental intervention might work to orchestrate new alignments between spatial relations and spatial zones. We analyse planning as a form of 'governmentality', one that employs key geographical ideas about the spatial realm. We investigate how planning – as a 'network of knowledge' – interacts with its environment, selectively drawing in some aspects of its surrounding context while excluding others. We see how this process of interaction leads planning to construct particular spatial imaginaries. In reviewing these imaginaries, we see that early forms of planning focused primarily upon the physical characteristics of places. Over time, however, social and political processes gained a higher profile. More recently planning has come to embrace not just social entities but natural entities as well. However, it is argued that planning struggles to accommodate these entities, in part because it is unable to adequately engage with spaces of heterogeneity and fluidity. Suggestions are therefore provided for some amendment to planning processes so that a more dynamic approach, oriented to the multiplicities of space, can be brought into being.

Some indications as to how planning might play this new role emerge in the final case-study chapter. Through the analysis of two competing food networks, it is shown how contemporary food spaces are forged in relational terms. Here we find networks of spatial simplicity confronting networks of spatial complexity. In the former, efforts are made to disseminate a uniform set of spatial relations so that spatial location and network come to resemble one another. In this network, food becomes a standardized input into the network-building process. In the latter network, spatial diversity is maintained so that only loose connections are established between varied food spaces. Because this second network aims to promote diversity in food, it is better able to root itself in the multiplicity of food space. It thus illustrates how heterogeneous relations might be established in practice. We conclude that the interaction between network and space in the food sector might reveal how a post-structuralist politics of nature could be conducted.

The case-study chapters focus on relations between nature and society. They therefore illustrate how post-structuralist theory can be brought to bear on this traditional area of geographical concern. In the final chapter, the relationship between post-structuralist theory and ecology is further explored. The chapter takes as its starting point Verena Andermatt Conley's (1997) observation that post-structuralism has always had close ties to ecology but the strength of these has never been fully appreciated or investigated. It is suggested that the connections between the two approaches be explored further so that the contribution of post-structuralism to pressing ecological problems can be ascertained (the case-study chapters give only rough guidance on this issue). However, some cautionary notes are sounded in this chapter. These refer mostly to the status of the 'human' in ecological and post-structuralist theory. In post-structuralist accounts, humans are displaced from the centre of the analysis and attention is focused upon relations of various kinds (as we shall see, social actors are 'decentred' into relations and can only act once relations are in some way 'centred'). Likewise, ecology sees humans as only one amongst many ecological entities (as Robert Crawford's poem 'Bio' presented at the beginning of the book makes wonderfully clear). The final chapter proposes that, despite the 'anti-' or 'post-'humanist dispositions of these theories, it is still necessary to retain some conception of human distinctiveness, even if human actions are thought to result from relational affiliations. The significance of the human, it is argued, derives from the fact that discourses, texts, arguments and other mechanisms of meaning generation are aimed at motivating humans to engage with the world – peoples and natures – in particular ways (again, some of these ways are revealed in the case-study chapters). If post-structuralism is to be politically effective (as many of its adherents assume it should be) then some notion of the human receiver of the post-structuralist message must be retained. Moreover, the case-study chapters show that an effective post-structuralist politics of spatiality should be concerned with the interaction between emergent process and territorial coherence and it should aim to 'shape' or 'steer' this interaction in ways that ensure an enhancement of ecological diversity and integrity.

5

Dis/Ordering space I: the case of nature

Spatiality, however constructed, simultaneously unifies and separates. (Harvey, 1996)

Introduction

Post-structuralism in geography focuses on the ways that dynamic and complex processes move through and across space, modifying spatial entities, recasting spatial relations. Following the material presented in previous chapters, it might be assumed that as societies become fragmented and striated by networks, so processes of spatial decomposition (referred to in Chapter 1 above) will generate increasingly complex topologies in which complexity and fluidity continually undermine simplicity and stability. Yet, in the last chapter, we began to see that topologies do not always displace topographies: at certain times and in certain places, topographical spatial formations can be consolidated *within* topological relations; reterritorialization inevitably follows deterritorialization. In other words, we should treat post-structuralist celebrations of the 'becomingness' of space rather cautiously for, as Allen (1999: 328) points out, 'we still live in a world of fenced-off territories and exclusions'.

In this first case-study chapter, we consider the relationship between complex relations of becoming and the consolidation of 'fenced-off territories'. In particular, the chapter seeks to identify how spatial classifications struggle to 'contain' heterogeneous relations. Thus, it looks in some detail at the zoning of space and it examines how demarcated zones interact with Deleuzian processes of becoming and emergence. It suggests that we might see in this interaction not just an intermingling of simplicity and complexity (the zone and the relation) but the consolidation of new emergent powers – that is, the act of division itself *guarantees* the construction of transgressive spatial relations. Thus, we cannot simply propose relational solutions: we need to think about the territorial implications of relational processes. Moreover, in certain circumstances, it might be appropriate to assert territoriality *over* relationality. Such a circumstance emerges towards the end of this chapter. In developing these themes, the analysis takes its cue from Bruno Latour's (1993, 1999) suggestion that, while social institutions use classificatory schemes to routinely separate out forms of socio-spatial practice, in fact such schemes function to generate ever-increasing numbers of 'hybrid' entities (see also Whatmore, 2002).

Latour (1993) suggests that the interaction between division and relation is a defining characteristic of modernity. In particular, he sees attempts to sharply distinguish 'nature' and 'society' as emblematic of a 'Modern Constitution', one that 'believes in the total separation of humans and nonhumans' (1993: 37). While Latour emphasizes that the distinction between 'nature' and 'society' comprises a key classificatory motif within modern society, he also suggests that the tension between (simple) classification and (complex) relation is becoming increasingly difficult to ignore. In his view, 'purification' proceeds hand in hand with 'translation': 'far from eliminating mediation, [modernity] has allowed this to expand' (1993: 41). Thus, as Lee and Stenner say during a commentary on Latour's work:

> Modernity in this account, is founded upon a moment of systematic misrecognition: we must speak as if nature and culture are clear and distinct realms but act as if they were not. We produce the modern world by mixing natural and cultural things into productive hybrids who can then promptly be ignored thanks to purifying tendencies of modern thought. (1999: 95)

An appreciation of the 'double movement' (purification *and* translation) gives rise to 'non-modernity', a social arena in which the failure of nature–society classifications in sifting out the world is increasingly recognized. Latourr's own writings, which question the salience of modernist dualisms, can be seen as an illustration of 'non-modern' thinking (see also Michael, 2000; Whatmore, 2002).

A fundamental manifestation of the tension between modernism and non-modernism is political ecology (Latour, 2004). In line with the 'non-modern' perspective, many environmentalists believe that the separation between the 'natural' and the 'social' will ultimately be undermined by ecological relations (at some point nature will 'act' back upon human society, thereby disrupting and amending economic and social relationships – see, for instance, Beck, 1992). One main function of the environmental movement, therefore, is to remind modern society that development inevitably binds humans and nonhumans more closely together within complex socio-natural assemblages. And yet, while environmentalism is attuned to the hybrid character of the modern world, it is also caught up in the dualistic presuppositions highlighted by Latour (1993), for many environmentalists cling to the belief that nature can ultimately be separated from society. Thus, the objective of much environmental action is not to more deeply embed human action and human society in heterogeneous or hybrid relations; it is instead to diminish the impact of this society on natural entities by protecting nature from human interference.[1] These two strands of environmental thinking display the modernist paradox identified by Latour (1993): on the one hand, all (economic, social and political) actions (including environmentalism) unfold within a 'hybridized' or 'ecological' society in which natural and social entities become ever more relentlessly intertwined; on the other hand, environmentalism proclaims the need to (re-)establish a clear division between the two realms, so that nature is more clearly demarcated from social influences.

In this chapter, we take up this paradoxical aspect of environmentalism by assessing how nature has been spatialized in the environmental movement. In particular, we investigate a classificatory system that has been instrumental in allocating 'nature' and 'society' differing spatial zones. Our example in this regard refers to the urbanization process and its impact on the countryside. Urbanization has been an almost continuous feature of modern development and concerns about its impact can be traced back to the earliest phases of environmentalism (see Lowe and Goyder, 1983; Eder, 1996; Macnaghten and Urry, 1998; Sutton, 2004). As urbanization has unfolded, so it has sprawled further and further outwards, thereby disrupting rural nature. In response, environmental groups have attempted to establish a clear classificatory division between the 'urban' and the 'rural' in order to limit the impact of the city on the surrounding natural environment. Yet, as the politics of division has worked to distinguish two spatial zones, transgressive relations have emerged that operate across any such spatial categorization. In this chapter, we describe some of the challenges that confront the environmental movement as it attempts to protect nature in the face of hybrid and heterogeneous processes of change.

We examine processes of division and transgression mainly in the context of one country – England – where the struggle to differentiate 'urban' and 'rural' zones has been particularly fraught. England is not only a restricted landmass, with one of the densest populations on earth, but it was also the first nation to industrialize and urbanize its economy and society. In the nineteenth century, it shifted from being a predominantly rural-agrarian society to being a predominantly urban-industrial society; as a consequence, urban areas grew rapidly and began to engulf their surroundings. Rural areas seemed vulnerable, and this apparent vulnerability gave rise to robust attempts to protect them from urban sprawl. At the same time, the countryside came to be portrayed as the main repository of 'nature'. Thus, efforts to distinguish urban and rural have, in England at least, been interpreted as efforts to distinguish 'nature' and 'society'.[2]

In the following sections, we first examine the role of preservationist thinking in the context of the environmental movement. As we shall see, though preservationism comprises an early form of environmentalism, it has remained a constant presence in the movement as a whole. Having shown that the urban–rural distinction is of fundamental importance to environmentalism, we then discuss how the divide between the two spatial zones has been put in place in the English national context. We describe how the planning system has played a key role in demarcating the two spatial zones and we consider how this role has been buttressed by the activities of environmentalists. We then go on to show how transgressive urban–rural relations have emerged in tandem with the establishment of a clear spatial division between urban (society) and rural (nature). These transgressive relations can be seen as affecting both rural and urban areas: rural nature becomes incorporated within new socio-economic formations that serve to redefine the significance of both rurality and nature, while urban society finds itself more deeply embedded in complex ecologies, ecologies that serve to undermine taken-for-granted notions of the city as a purely social and economic zone somehow separated off from the natural world. We will conclude by re-assessing the status of spatial classifications and the way these might to be aligned with heterogeneous relations.

Demarcating spaces of 'nature' and 'society'

In Latour's (2004: 18) view, environmentalism has tended to affirm a conservative conception of nature: as he puts it, 'most of the time [environmentalism] changes nothing at all; it merely rehashes the Modern Constitution of a two-house politics in which one house is called politics and the other [...] nature'. In making this complaint, Latour is suggesting that environmental and ecological groupings have invested effort in upholding a profoundly 'modernist' distinction between nature and society. In a similar vein, Klaus Eder (1993, 1996) believes social views of nature have long fallen into two main 'camps': nature as a resource for human exploitation and nature as the source of ultimate goodness. This 'double structure', Eder argues,

> has its origin in the everyday practices that determine the interaction with nature [...] The ordinary practical basis of the double significance is seen in the dichotomy of city and country. The double symbolisation of nature enters into the antagonism between cultivated land and wilderness. It produces the antagonism between dominance and protection of nature, and it produces the peculiar relationship to animals that is torn between meat and mercy. (1996: 147)

Given the resilience of the 'double structure', it is perhaps not surprising to find it in some of the earliest manifestations of environmentalism. Importantly, in the English context this double structure comes to be seen in terms of an urban–rural separation, with nature and society allocated to differing spatial zones (Williams, 1973). For instance, Keith Thomas (1984: 301) shows that the desire to separate nature from society stems from the eighteenth and nineteenth centuries when 'the growth of the towns led to a new longing for the countryside. The progress of cultivation had fostered a taste for weeds, mountains and unsubdued nature'. As urban areas began to grow in the wake of industrial advance, writers such as Wordsworth and Ruskin focused their emerging environmental aspirations on the Lake District and other areas of 'unsullied' nature. For these early campaigners, 'nature' was deemed to lie far from industrial England: it 'came in a sense to be cast out of such urban-industrial spaces and to find its "home" on the very margins of the emerging industrial society, in parts of the British countryside' (Macnaghten and Urry, 1998: 175). Moreover, this ruralized nature was no longer viewed as robust but as vulnerable, threatened by urban growth and industrial expansion. The rapid and sprawling character of urban areas in the later years of the nineteenth century seemingly heightened the threat.[3]

Growing concern about urban encroachment on rural nature led directly into the formation of 'preservationist' organizations at the turn of the twentieth century, including the Lake District Defense Society in 1883, the Royal Society for the Protection of Birds in 1891, the National Trust in 1895, the Society for the Promotion of Nature Reserves in 1912 and the Council for the Preservation of Rural England (CPRE) in 1926 (Lowe and Goyder, 1983). In seeking to combat urban sprawl, these agencies mobilized ideas associated with a threatened rural nature and they sought to 'represent' nature in political disputes over patterns of development and regulatory responses to those patterns. The main concern was 'urban sprawl'. It was assumed that this pernicious process could only be restrained by concerted government action, notably through the establishment of a comprehensive and robust land use planning system. Yet, despite vigorous campaigning by the movement's elite members, little in the way of preservationist planning emerged in the early years of the twentieth century. Although the urban environment was improved through public health legislation, urban sprawl continued to unfold, largely as a result of increasing car ownership and the development of trunk roads (Clapson, 2000). Lines of flight from the city ran further and further into the countryside thereby challenging nature's distinctive status and its integrity.

These trends were perceived in highly negative terms by the preservationists, as Thomas Sharp, a planning theorist and leading CPRE member, makes clear:

> From dreary towns, the broad, mechanical, noisy main roads run out between ribbons of tawdry houses, disorderly refreshment shacks and vile, untidy garages. The old trees and hedgerows that bordered them a few years ago have given place to concrete posts and avenues of telegraph poles, to hoardings and enamel advertisement signs. Over great areas there is no longer any country bordering the main roads: there is only a negative, semi-suburbia. (1932: 4)

Sharp illustrates here how heterogeneous processes of change run headlong into clearly ordered spatial zones: the urban destroys the rural by generating hybrid entities. For Sharp, the only solution was a new zoning system which allocated entities and activities into discrete and clearly differentiated spatial areas. As Patrick Abercrombie (1933: 36), another planning theorist and founder member of CPRE, famously put it: 'the essence of the aesthetic of the Town and Country Planning system consists in the frank recognition of these two opposites [...] Let Urbanism prevail and predominate in the Town and let the Country remain rural. Keep the distinction clear'. In David Matless's (1998: 51) view, Abercrombie was expressing here 'a particular form of modernism', one that subscribes to 'orderly progress through planning'. City and countryside were to be governed as distinct and opposed geographical spaces:

FIGURE 5.1 Contemporary concerns about the 'hybrid' character of the countryside are exemplified in the CPRE's 'cluttered countryside' campaign (Source: CPRE, 1996, reproduced courtesy of the Campaign to Protect Rural England)

'a normative geography of distinct urbanity and rurality [was to be] asserted over an England-in-between of suburb, plotland and ribbon development' (Matless, 1998: 32). In other words, hybridity and heterogeneity should be firmly kept at bay.

Yet, despite the energetic activities of preservationist groups, sprawl continued. During the inter-war period, over four million new houses were constructed in England: as the planning historian Ward (1994: 43) points out, 'the overwhelming majority of these were suburban houses, usually in semi-detached or short-terraced form, built at densities of twelve to the acre or less, with great emphasis on private gardens [...] a vision which became strongly associated with the garden housing of the new suburbs'. Not surprisingly, preservationists feared that suburban growth – especially when it straddled the ever-increasing numbers of new roads – would ultimately destroy the remaining repositories of nature to be found in the countryside.

Modernist planning only found favour with government following World War Two, when strategic concerns to boost food production gave new urgency to efforts aimed at protecting rural land (Sheail, 2002). At this time a comprehensive planning system was put in place by the 1947 Town and Country Planning Act, a piece of legislation that gave official sanction to the establishment of an urban–rural divide.[4] To achieve this separation, the new planning system assigned planning functions to the city authorities and administrative counties, to be exercised under the supervision of a planning ministry at the national level, which would issue directives to ensure standardized procedures across the national territory. In organizational terms, this distribution of responsibilities ensured a comprehensive governmentalization of planning. However, it also introduced a sharp division between urban and rural areas as the urban boroughs and rural counties became clearly separated from one another. On each side of the divide, it was envisaged that natural and social entities would be assembled in carefully coordinated topographical formations. In the urban zone, dynamic processes of change would be encouraged in the hope that robust economic formations would come into being. In the rural zone, the only economic activity given any legitimacy was food production – in the minds of preservationists, agriculture was working with nature and could therefore be regarded as somehow 'natural' (Green, 2002).

This spatial 'settlement' was bolstered by the multitude of preservationist groups that could be found across rural England. During the 1960s and 1970s, there was a marked increase in environmental activism at the local level and local planning agencies found themselves embedded within dense networks of local preservationist groupings.[5] The main aim of these groupings was to ensure that local planning agencies adopted preservationist governmentalities in their decision-making processes. In their view, topographically robust spatial formations would keep topological complexity at bay.

Yet, just as the preservationists succeeded in placing nature within its own protected zone in the countryside, environmental problems began to emerge that seemingly disrupted the new spatial settlement. Macnaghten and Urry highlight this potential disruption when they say:

> By 1970 public attention, both in Britain and abroad, began to be drawn to a much wider range of problems threatening the environment, concerns not simply over wildlife conservation and amenity, but now including nuclear radiation, pesticide use, vehicle emissions and other systemic forms of air and water pollution. These events began to generate an awakening sense of a more general crisis of environmental bads, moving across national borders and potentially invading everyone's body. (1998: 50)

The new concerns appeared to run across all spatial distinctions such as that between urban and rural areas. Moreover, these 'transgressive' environmental issues were articulated by new environmental groups, such Greenpeace and Friends of the Earth (FoE). The rise to prominence of the new groups appeared to represent a challenge to preservationism and its zoning approach to environmental problems.[6] Environmental issues were now redefined as 'ecological' problems. In proposing this redefinition, the new groups indicated that nature should no longer be seen as simple and static but rather as dynamic and complex. Its protection could not be assured using simplified spatial classifications: now, holistic, all-embracing approaches were needed so that the integrity of natural systems could be assured.

Yet, despite the attention given to ecological issues, preservationism continued to hold its own. As Sutton (2004) points out, a major part of the growth in membership in environmental organizations during this latest phase of campaigning was in preservationist groups. As he (2004: 45) says, 'the majority of new members drawn into the environmental movement joined either moderate groups and organizations or, more significantly, organizations that predate the new environmentalism'. Thus, we discern some continuity in the organizational structure of the environmental movement, with the older groups playing a key role in generating a mass membership in the later years of the twentieth century. Moreover, the new members of the older groups held the same preservationist aspirations as the older members, that is, they retained a concern to find stable spatial zones for nature in the countryside, nature reserves, sites of special scientific interest and so forth.[7] Thus, although the recent period has seen the politics of preservationism encompassed within much broader environmental issues, it remains an important part of the movement's repertoire. The environmental movement now seeks to integrate a focus on 'relational' or 'ecological' natures' with its traditional reliance on 'zoned' natures.

BOX 5.1

The spatial imaginary of preservationist environmentalism in England:

- A key strand of environmentalism emerged during the last years of the nineteenth century and the early years of the twentieth century, which saw the main repositories of nature as lying beyond the towns and cities in the countryside.
- This nature was fragile and fugitive, threatened by ever-encroaching urban sprawl. The urban threatened ruralized nature because it brought the 'polluted' and 'degraded' city into areas that had previously maintained a balance between human ways of life and natural processes.
- The solution to this threat was the establishment of a divide which would keep heterogeneous urban processes at bay and would keep rural nature safe and secure. The divide would be policed by state planning authorities, working in line with preservationist governmentalities.

To conclude this section, we can say that the evolution of the English environmental movement indicates that as industrial growth modifies the environment, political actors continuously work to diminish the effects of any such modifications. The primary means of limiting industry's impact becomes, at least in England, a division between urban and rural, with industry and society allocated to the urban realm and with nature residing in the rural. In short, nature is spatially zoned *out* of society. Although in more recent times environmentalists have turned to examine complex ecologies of dynamic and emergent relations, the zoning approach remains salient. There is still a strong desire to demarcate places of nature from spaces of society.

Yet, demarcations of this kind immediately give rise to a paradox: industrial society leads to greater and greater interrelationships between productive activity and natural resources; at the same time preservationist movements believe a 'line' or a 'divide' should be drawn between industrial society and its environment. In order to make the spatial demarcation stick, a series of 'transgressive' relations need to be curtailed. That is, a purification of space needs to be undertaken to ensure that the spatial classifications used by preservationist planning and other regulatory mechanisms correspond to the collections of entities to be found in the two clearly separated spatial areas of 'urban' and 'rural'. In the next two sections, we will examine this paradox a little more closely by considering how the two zones have fared since they became divided one from the other. In presenting this analysis, we will argue not only that spatial transgressions inevitably problematize the strict division between the two zones but also that the interaction between division and relation generates new spatial forms. In short, while spatial classification fails to contain spatial relation, the encounter between the two stimulates the emergence of new spatial assemblages.

Rural transformations of (rural) nature

We begin on the rural side of the divide. As we have seen above, the preservationist movement assumed that, if the countryside could be protected from the expansion of the urban then nature would be left free to flourish. This view was based on the (romantic) notion that natural landscapes are rural in character: even though these landscapes have been modified by agriculture and other land-based industries, rural society remains the best custodian of nature (echoes of this preservationist assumption can be found in Scruton, 2004). This rather simplistic view has long carried considerable weight in the preservationist movement and, as preservationist pressure groups came to influence legislation in the post-war period, so a rather crude distinction between rural nature and urban society was enforced. Indeed, the post-war planning system was premised on the notion that if the urban could be contained within its pre-existing boundaries then rural nature would endure (Hall et al., 1973). However, in making this assumption the preservationists also assumed that nature could co-exist with the dominant economic activity in rural areas (in terms of land use) – agriculture.[8] Lowe et al. summarize preservationist views of agriculture as follows:

> Farming practices seemed to pose no possible threat to other rural interests and pursuits. On the contrary, it was felt that the debilitated condition of farming exacerbated many

other threats to the countryside, such as urban encroachment and the decline of rural communities. A secure and revitalised agriculture was seen as the essential conserver of both rural life and the natural beauty of the countryside. (1997: 2)

Yet at exactly the same moment as the 1947 Town and Country Planning Act was putting in place safeguards for the protection of agricultural land, the 1947 Agriculture Act was assisting the agricultural industry to become a fully-mechanized form of manufacturing industry. Under this measure, the state was required to initiate the wholesale governmentalization of agriculture. This governmentalization process was based on a series of measures that were aimed at the rationalization of the agricultural industry:

1. The main role of government was to administer a system of 'guaranteed prices' so that farmers would be paid for whatever they produced, irrespective of market demand. One main effect of this system was to reward those producers whose production was greatest. Thus, the largest farms gained the most in terms of financial support. As a consequence they became large, intensive farms.
2. The state sought to increase agricultural efficiency and competitiveness through the provision of grants that encouraged farmers to undertake land development initiatives and to increase their levels of mechanization. Thus, the large intensive farms also became technologically sophisticated in their farming practices.
3. Farmers were encouraged to adopt common business and husbandry practices by state extension agencies – that is, business practices were subject to processes of governmentalization in which standardized accounting and other procedures were disseminated. The aim was to ensure enhanced standards of economic efficiency on farms..

All these measures were aimed at turning agriculture into an efficient and productive industry. They also sought to transform farmers into innovators and entrepreneurs. In order to achieve the latter aspiration, a Panoptical regime of agricultural regulation was brought into being, in which the state 'micromanaged' the practices of individual farmers in line with a particular set of productivist governmentalities (Murdoch and Ward, 1997). The state became involved in almost all aspects of agriculture from the provision of research and development to the monitoring of on-farm business and husbandry practices. The effectiveness of this regime can be seen from the vast increases in production that took place during the early post-war period: by 1969 agricultural output stood at nearly twice its pre-war level while the number of farmworkers employed on farms more than halved during the same period. In short, the entire agricultural industry moved some considerable way to becoming a modern and efficient industrial system – that is, there was a vast increase in the amount of technologically sophisticated machinery on England's farms and ever-greater amounts of pesticides and chemical fertilizers were spread across the

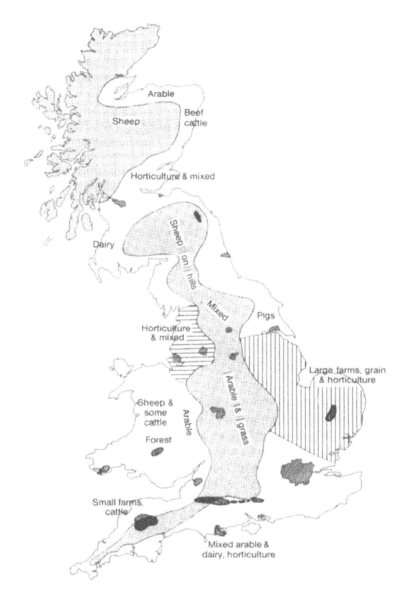

FIGURE 5.2 In the post-war period, a 'national farm' was 'governmentalized' with differing regions specializing in differing commodities (Source: Norton-Taylor, 1982)

land. These material outcomes followed directly from the productivist governmentalities that lay at the heart of state agricultural policy.

The interaction between government regulation and growing technological development in the agricultural sector meant that England's rural areas became part of a micro-managed, spatially coordinated, topographical zone (a so-called 'national farm' – see Murdoch and Ward, 1997). The management and co-ordination of this zone was aimed at increasing food output: thus, farm structures, agricultural working practices, new machinic assemblages, all were to be mobilized in pursuit of this objective. In the process however natural entities were transformed, neglected and destroyed. Hodge summarizes the changes wrought by the post-war agricultural regime as follows.

> In the 40 years following the Second World War, about 95% of lowland meadow was lost, 80% of chalk downland, 60% of lowland bogs, 50% of lowland marsh and 40% of lowland heath [...] The length of hedgerows declined from 495,000 miles in 1947 to 386,000 in 1985 [...] There is also continuing concern at the loss of wildlife, recent studies emphasizing that this is an indirect consequence of pesticides which have killed important food sources in intensively farmed areas. (Hodge, 2000: 103)

Agriculture was progressively being 'lifted out' of its natural resource base and as this 'lifting process' ensued so rural nature was transformed. It therefore became increasingly difficult to sustain the preservationist assumption that agriculture somehow 'worked with' nature; rather it seemed to be working against it in profoundly damaging ways (for a full account of the damage, see Harvey, 1997).

As the impact of the post-war revolution in agriculture became apparent, so preservationist groups belatedly began to shift their attention from urban pressures on the countryside to the destructive effects of rationalized farming practices. In making this shift, they initially campaigned for changes in productivist governmentalities so that impacts on nature gained a more central place in government thinking. However, the preservationists made little headway in this regard, in part because the agricultural policy community was extremely effective at excluding everyone but farmers from key policy arenas (Winter, 1996). The preservationists thus adopted another tactic: they began to suggest that agriculture should be incorporated into the land-use planning system so that any new development of farmland would be assessed by local planning authorities. The aim here was to bring agriculture into local policy arenas dominated by preservationist political networks (Lowe, 1977). Yet, this brings us to another paradox – local policy arenas were only dominated by such networks because so many preservationist activists were now present in rural areas. And these activists were present in rural areas because a fundamental shift was taking place in rural society – in particular, there was a sharp increase in the number of ex-urban rural residents living in the countryside. Moreover, the movement of population from urban to rural was accompanied by a movement of industry and services, meaning that the economic activities present in the countryside were no longer dominated by agriculture.[9] In other words, dynamic processes of socio-economic transformation had drawn urban and rural areas more closely together. The imposition of spatial zones had facilitated the emergence of heterogeneous relations that combined the urban and the rural in new ways (for instance, many people now lived in the countryside but worked in the city ensuring continuous flows of population across the urban–rural divide).

BOX 5.2

The rural transformation of (rural) nature stems from:

- The state's adoption of a national agricultural policy in the post-war period. This policy led to a full governmentalization of the agricultural sector with the state micro-managing farming practice in line with (Foucaultian) strategies of normalization.
- The aim of the policy was increased food supplies. Thus, a heterogeneous assemblage of resources was harnessed in the line with this policy goal: new farming structures, more machines on farms and fewer workers, more pesticides, more fertilizers, bigger fields and so forth.
- The consequence was a denuded nature and the construction of a simplified rural topography in which rural space was constructed around the needs of a productivist agriculture: bigger fields, polluted water, fewer hedges, less wildlife and so forth.

At first sight, these 'transgressive' processes would seem to fatally undermine the post-war spatial settlement. However, Hall et al. point out that:

> The majority of English villagers [...] are adventitious to the countryside. They are either longer-distance commuters to the towns, or retired people [...] They tend to be prosperous and well-organised, and they care a great deal about the countryside and the way of life it represents. They see the countryside as a repository of tradition and of stability in the face of change. They naturally wish to preserve this image which makes them profoundly and instinctively conservative or conservationist – the two words in this context are synonymous. (1973: 431)

Not surprisingly, as this social group came to dominate rural society, so membership of local preservationist societies increased.[10] Thus, farmers found themselves encompassed within a new social formation, one that was quite unsympathetic to the governmentalities at work in agricultural policy. As Lowe et al. say during a study of agricultural pollution in the county of Devon in South-West England:

> The influx of large numbers of newcomers was [...] associated with, and helped catalyse, a major shift in public attitudes to agriculture and the countryside. Many farmers had new neighbours with quite different perceptions of the function of the countryside. [They now] experienced direct pressure from neighbours and local people to change their farming practices. (1997: 155)

Local environmental groups were able to mobilize the new, ex-urban social formation in pressing for stronger local policing of polluting agricultural activities – that is, the groups ran campaigns alerting local residents to the farm pollution 'threat' and encouraged concerned citizens to report any transgressions to the relevant policing authorities. As a result of this socially embedded form of pollution regulation, farmers soon came to feel themselves 'under siege' (Lowe et al., 1997). Rural nature therefore came to be preserved not by actors and institutions that were indigenous to rural space but by urban social groups that had moved *across* the urban–rural divide.

Urban transformations of rural nature

Despite the greater attention given to agriculture, however, the activities of the newly strengthened preservationist groups were mainly focused on planning. As we saw earlier in the chapter, local environmental groups came into existence to ensure that local authority planning subscribed to the preservationist prin- ciples that lie at the heart of the planning system (summed up in the couplet 'rural preservation/urban containment', Hall et al., 1973). In the wake of the urban–rural shift of population and industry, the countryside was once again perceived as being under threat from lines of flight out of the city. The main threat was seen now to be the new house building that facilitated population flows across space. As the movement across the divide intensified, it began to run up against traditional preservationist concerns about the loss of land (and thus rural nature) to housing development. This conflict became especially acute in the south of England where the pressure for new homes had long been intense, especially in the 'well-heeled' shire counties to the south, west and north of London (Short et al., 1986; Murdoch and Marsden, 1994). The demands on rural land for housing emerged in this region because of a drift of population from north to south and a simultaneous movement of households away from the cities. It was therefore calculated that the population of the south-east region would continue to grow, but most importantly this growth would be concentrated in the areas *outside* London (Allen et al., 1998).

Concerns about the suburbanization of the countryside reached their height in the mid-1990s, when the Department of the Environment published hous- ing projections which forecast that the number of households in England would grow by 23 per cent over the next twenty years. It was calculated that around half this total would be located outside the major conurbations with the consequence that rural areas, especially those in the south-east region, would face another wave of acute housing pressure (see Breheny, 1999; Vigar et al., 2000; Murdoch and Abram, 2002). Following the publication of this figure, preservationist groups such as the CPRE launched a major campaign aimed at preventing a despoilation of rural nature by further rounds of urban- inspired housing development. This campaign was conducted at both the national and local levels and generated considerable concern in the rural areas of southern England. In response, the government began to suggest that per- haps a large proportion of the new houses should be located in urban areas. In so doing, the government once again asserted the urban–rural divide as the best means of protecting rural nature.

In order to explore ways of levering these houses into urban locations, the government established an Urban Task Force to be headed by the architect Richard Rogers. From the Task Force report, published in 1999, it seemed that a concern for the relationship between urban and rural lay at the heart of the group's deliberations. The threats to (rural) nature were all seen to emerge from within the urban realm as destructive processes of change were continually being re-generated from within this spatial zone. Thus, implicit in the Task Force's proposals was a strengthening of urban–rural distinctions. For instance, after outlining the environmental problems that follow from urban sprawl (for example, increased energy use associated with low-density housing and car-dependent travel patterns), the report said:

> Ultimately, town and country are interdependent. The welfare of one cannot be secured at the expense of the other. The guiding principle must be, therefore, that we focus maximum efforts on using available building land within our existing urban fabric. This does not mean that there will be no new greenfield development or that some of that development will not intrude upon existing green belts. What is important is that where such development has to take place it is based on strong principles of sustainable urban design, and it minimises its impact upon the surrounding countryside. (Urban Task Force, 1999: 37)

Many of the concrete proposals put forward by the Task Force followed from this perceived need to constrain urban sprawl. Moreover, the report seemed to have a profound influence upon government policy: in a policy guidance note for planning, published in the 2000, the government urged the planning profession to promote 'sustainable patterns of development'. However, 'sustainability' was defined here as the concentration of most new housing development in urban areas (DETR, 2000: 1).

BOX 5.3

The continuation of urban sprawl in the 1990s led to:

- A renewed focus on the zoning of urban–rural, nature and society so that countryside locations could be protected against urban housing.
- A focus on the need for an 'urban renaissance' so urban processes and entities could be better retained within urban locations. This urban renaissance would attend to the complex ecologies of urban life in the hope that transgressive relations across the urban–rural divide could be constrained.
- The paradoxical result that rural preservationists found themselves pulled across the divide into the urban realm in order to argue for enhanced urban environments. The divide was now set within a complex ecological context.

Although the Urban Task Force report and the government's 'urban renaissance' policy proposals seem at face value to be a triumph for preservationism, the thrust of the new approach paradoxically pulled the preservationist movement in a new direction, for it now seemed that *rural nature* would be best protected by improvements in the environmental quality of *urban areas*. Thus, local activists began to lobby for urban housing capacity studies, better standards of urban housing design and 'sustainable' urban extensions (Murdoch and Lowe, 2003). The protection of rural nature came to be seen as only one part of a much larger parcel of 'goods' (improved urban environments, environmentally benign patterns of living and better use of scarce resources, etc.) that could be delivered through preservationist governmentalities. In this regard, the preservationists seemed to be aligning their traditional concern for urban–rural division with the recognition that both urban and rural areas should be combined within complex relational ecologies, as social and industrial changes draw differing spatial zones into states of interdependency. In such states of interdependency, the assertion of spatial divisions needs to go hand in hand with the assertion of spatial relations, an outcome that makes it hard for preservationists to focus solely upon on a separated and preserved nature.

Urban transformations of urban (and rural) nature

This recognition of the need to environmentally manage urban and rural areas in tandem with one another brings us inevitably to the natural qualities of urban areas. As we have seen above, the environmental movement has generally assumed that in England, at least, nature is to be found in the countryside while society is to be found in the city. Thus, it is usually thought that efforts to protect nature should be focused on preserving the rural and containing the urban. Yet, it is clear that nature is not easily sifted into any such spatial division. As Chris Philo (1998) illustrates in his description of Smithfield market in London, nature has always been present in the city. This can be seen, he says, in the number of animals kept in the city as pets, as zoo species and as livestock for slaughtering. In discussing the way animals disrupt our taken-for-granted spatial orders, Philo quotes Atkins, who says:

> The idea of finding animal husbandry in an English city in the present or in the past might appear strange in view of the current pressure of urbanisation upon agricultural land use. The built-up area somehow seems an alien environment in which to keep horses, cows, pigs and sheep, but in mid-nineteenth-century London the idea of a clear-cut distinction between urban and rural life had yet to develop. (Philo, 1998: 59)

Yet, even once such a divide was established in the twentieth century, animals continued to inhabit the city, with pets, zoo animals, wild animals and livestock still very much in evidence (Wolch, 2002).

Nature is present in the city in many other forms as well – in well-tended parklands and gardens, in woodlands and in neglected or 'wasted' spaces.[11] Again, there has been a long-standing concern for urban nature, as Laurie explains:

> The development of parks and the preservation of natural amenity in nineteenth-century cities was the result of a movement of great strength and persuasion. This movement was built on five concepts. First, that natural or natural-looking parks, street trees, and public gardens would improve the health of the people by providing space for exercise and relaxation in pure air. Secondly, it was believed that the opportunity to contemplate nature which public parks provided would contribute to a much needed improvement in morals. Thirdly, a fascination with the aesthetics of natural landscape in the second half of the nineteenth century led to the notion that parks and gardens would improve the appearance of a city. Fourthly, and in association with this, the value of property would be increased due to its association with parks. Fifthly, an increasing public interest in natural processes and the elements of nature, both plants and animals, fostered the introduction of educational arboreta and zoological gardens and contributed to the desire for natural areas with indigenous plants as habitats for wild life. (1979: 37–8)

The confluence of these various factors led to the emergence of a political movement dedicated to the protection of urban green space. The movement can be traced from the formation of the Commons Preservation Society in 1865, through to the Open Spaces Society established in 1893, and on to the various urban trusts that appeared in the post-war period. Such groups have acted to protect urban green space against further development and have attempted to regenerate forgotten and neglected spaces so that urban biodiversity might be increased.[12] As Whatmore and Hinchliffe (2003: 1) emphasize, 'urban wildlife groups, amateur naturalists, voluntary organisations and the like have been key players in [the] realignment of urban spaces and conservation concerns'. Thanks to urban environmental activists, nature remains an ever-present feature of urban life.[13]

The existence of nature in the city can lead to new evaluations of green urban living. As Jim (2004: 311) puts it, 'a city with high-quality and generous green spaces epitomises good planning and management, a healthy environment for humans, vegetation and wildlife populations and bestows pride on its citizenry and government'. Yet, preservationism threatens these high quality natural environments as it extols the virtues of the compact or contained city. Jim admits that,

> Compact urban areas are characterised by the close juxtaposition of buildings and roads with limited interstitial spaces to insert greenery [...] The compact city here encompasses the high density built form with a high proportion of the land surface covered by buildings and other artifical structures and surfaces. The ratio of impervious to pervious land areas is very high and conditions for plant and animal life are usually very trying. (2004: 312)

The danger arises that, within the governmentality of 'urban compaction', nature can often become little more than an afterthought. In Whatmore and Hinchliffe's

(2003: 4) view, planners in the compact city rarely see 'the fecund world of creatures and plants as active agents in the making of environments'. Natural processes are still regarded as lying 'firmly outside the city'; the 'feral spaces in the city that sustain them are cast as "wastelands" ripe for development'.

Whatmore and Hinchliffe believe this problematization of urban nature is implicitly invoked in the report of the Urban Task Force (1999), mentioned at the end of the previous section. The main objective of the Urban Task Force report is to bring well-designed but contained cities into being. One key means of achieving this is thought to be an increased density of housing. The Task Force (1999: 64) thus recommends what it calls 'pyramids of intensity', which would facilitate 'intense and integrated development' on many so-called 'brownfield' (that is, previously developed but abandoned) sites. In order for these 'pyramids of intensity' to come into being, the planning system must act to increase housing densities in its plans and development-control decisions. Moreover, local authorities should 'undertake regular physical surveys of sites' to assess the potential contribution 'brownfield' land can make to meeting housing needs (1999: 214). These so-called 'urban capacity' studies can be used to explore land-development areas that would otherwise be neglected. Such surveys and studies should be used to prioritize governmentalities of urban renewal and regeneration (Murdoch, 2004). The consequence, as Whatmore and Hinchliffe note, is that urban spaces of nature will be squeezed even further so that 'brownfield' sites that include natural habitats will be released for development.

While the 'squeezing' of nature inside cities is the inevitable consequence of sustained but contained urban growth, the main urban pressure on the natural world can be attributed to the 'lines of flight' in and out of cities. Giradet describes these 'lines' in the following terms:

> The metabolism of most modern cities [...] is essentially linear, with resources being 'pumped' through the urban system without much concern about their origin or about the destination of wastes, resulting in the discharge of vast amounts of waste products incompatible with natural systems [...] Food is imported into cities, consumed, and discharged as sewage into rivers and coastal waters. Raw materials are extracted from nature, combined and processed into consumer goods that ultimately end up as rubbish which can't be beneficially reabsorbed into the natural world. More often than not, wastes end up in some landfill site where organic materials are mixed indiscriminately with metals, plastics, glass and poisonous residues. (1999: 10)

The flows of resources and wastes are orchestrated by networks which work to transform (rural) nature into discrete forms that can then be transported into the urban realm for the benefit of urban consumers. Once consumed, this nature gives rise to wastes that must then be exported back out to retain the 'purity' of urban space. In the process the city progressively transforms its external environment. Thus, the urban–rural divide not only fails to prevent the flow of materials in and out of the city but ensures the rural remains a zone ripe for exploitation.

The impact of the city on its external environment has been termed the urban or ecological 'footprint' (Wackernagel and Rees, 1996). In order to

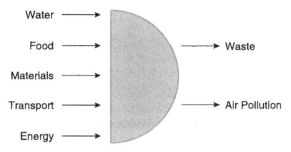

FIGURE 5.3   Heterogeneous flows through the city (Source: Urban Task Force, 1999)

calculate the 'footprint', flows are converted into the areas of land (or sea) required to deliver the requisite volume of materials. On this basis, it has been calculated that London's ecological footprint extends to the size of the UK, 250 times larger than the city's geographical area (Environment Agency, 2002). Such a finding indicates that even when a clear division between urban and rural is established, the flows of materials continue. In fact, they not only continue, they also increase. For instance, we noted earlier that as rural England was zoned as an area of nature to be sustained by agricultural production, the amount of food that could be produced for urban consumers rose markedly. In other words, the flow of natural resources into English cities from the English countryside (in the form of food) was enhanced during a period when rural nature was seemingly under sustained governmental protection.

BOX 5.4

The concern for urban nature drew attention to:

- The fact that urban areas had always been repositories of nature. Animals had long been kept in the city. Parks and other areas of outdoor amenity had always played a key role in urban life.
- That despite the number of organizations working to protect urban nature, the prevalence of the zoning governmentality in policy circles often led to the disregard for urban natural entities in development. This became especially clear in arguments over 'urban renaissance', for the new focus on urban development seemed to threaten urban nature anew.
- The fact that nature must be seen not in spatial but in ecological terms. Heterogeneous processes flow across spatial divides and carry significant consequences for natural entities in both urban and rural locations. The two spatial zones should thus be addressed simultaneously.

It is clear, then, that the main ecological effects of cities cannot be mitigated by simply dividing the urban from the rural: rather, an ecological approach is required that addresses both sides of the divide simultaneously. As Girardet puts it:

> Cities will need to adopt circular metabolic systems to assure their own long-term viability and that of the rural environments on whose sustained productivity they depend. To improve the urban metabolism, and to reduce the ecological footprint of cities, the application of ecological systems thinking needs to become prominent on the urban agenda. Outputs will also need to be inputs into the production system, with routine recycling of paper, metals, plastic and glass, and the conversion of organic materials, including sewage, into compost, returning plant nutrients back to the farmland that feeds cities. (1999: 10)

This ecological approach to urban–rural relations focuses on flows, on the heterogeneous materials that run across space in and out of the urban realm. Thus, we discern an ecological politics focused on the consolidation of a dynamic and complex system of socio–natural relations in which the urban and the rural are combined in some kind of 'sustainable assemblage'. However, it is important to note that within such an assemblage, zoning still remains necessary. As Girardet (1993: 156) points out, 'cities need to protect the farmland, forests and watersheds in their vicinity'. They therefore need to establish 'circular' relations in which the valued natural assets of the city and the countryside are nurtured and sustained simultaneously.

Conclusion

In this chapter, we have seen the emergence of a division between 'nature' and 'society' in the form of a spatial classification of urban and rural areas. It has been suggested that environmental groups have spent much of the twentieth century engaged in a politics of spatial division. The aim of this politics was to ensure that society was contained within the city while nature was protected from urban influence. This objective was successfully enshrined within dominant governmentalities, notably the 1947 Town and Country Planning Act, a measure that largely functioned to enclose the city and to preserve the countryside, and the 1947 Agriculture Act, which sought to ensure that rural nature was encompassed within a Panoptical regime of agricultural governance. A zoning governmentality thus came to prevail in the governance of nature.

One thing we might assume following this account is that social actors routinely focus their attentions on topographically simplified spatial zones. The idea that we can solve socio-ecological problems through some form of spatial separation remains powerful in the minds of the most engaged political actors. However, as Latour (1993) emphasizes, it is becoming more and more apparent that behind any such spatial divisions lie relations. Thus, despite the separation of urban and rural areas into dominant political arenas, transgressive processes continue to operate 'on the ground'. Moreover, again as Latour emphasizes, these transgressive processes are actually strengthened by the imposition of classificatory regimes. By preserving 'rural nature', the governmentalities of planning ensure the enhanced attractiveness of the countryside to urban consumers: population change straddles urban and rural areas, while increased food production (for urban consumers) transforms rural space. Preservationism is therefore confronted with a paradox: it supports the separation of the urban and the rural in order to preserve the latter; yet the implementation of preservationist policy ensures the generation of transgressive processes of change.

As we have seen, one response to this paradox is to set the urban–rural divide in an 'ecological' context. Thus, preservationism is forced to move away from environmental simplicity towards ecological complexity with the consequence that the connections between urban and rural areas become more important than the divisions. The aim of an ecological approach is therefore to bring urban and rural areas into a 'sustainable' alignment, one that opens up a possibility for concerted action to protect not just vulnerable rural natures, but neglected urban environments also. Ideally, this alignment should work to establish new and robust connections between spatial zones that have for too long been distanced from one another. In short, simplified spatial divisions should now be recast so that they can be encompassed within broader 'sustainable assemblages'. These assemblages should comprise rich ecologies of the human and the non-human, the social and the natural, the material and the immaterial. They will serve to link previously divided spatial zones into complex sets of spatial relations. In the spaces of these relations, differing mixtures of entities will be discerned so that some semblance of 'urbanity' and 'rurality' remains. However, such zones will no longer be seen as 'pure' for the interaction between spatial division and spatial relation will continue to generate new, hybridized spatial forms (Whatmore, 2002).

Thus, the 'politics of zoning', which has tended to dominate environmental politics, gives way to a 'politics of becoming', in which innovative and creative alignments take precedence. Here, new collectives are orchestrated so that heterogeneity and sustainability are achieved simultaneously. However, it also appears that topographical zones will still need to be successfully combined with topological processes. In other words, clearly coordinated sets of (environmental) entities will need to be established in ways that are sensitive to (ecological) processes of becoming and emergence. How this combination of topographical management and topological fluidity might be achieved is currently an open question. However, the case study presented above does indicate that zones and relations must be made fundamental to any strategy aimed at sustaining the heterogeneous ecology of space and place. Zones depend on relations and relations emerge from zones. A fuller recognition of this fact might aid the development of governmental approaches that move beyond two opposed spatial forms into a new spatial dimension, in which discrete areas are defined both by what they have been and what they will become, by their urbanity *and* their rurality, by the quality and diversity of their natures, by the ecological sensitivity of their socialities.

SUMMARY

In the main, this chapter has addressed the relationship between dynamic processes of becoming and spatial contexts of territoriality. It has done this by looking at efforts to protect and preserve nature using spatial divisions and designations in the context of England during the twentieth century. Following Bruno Latour, it has been argued that efforts to establish spatial divisions will inevitably be undercut by the dynamic nature of spatial relations. Latour's observation has been borne out by the simultaneous emergence of divisions and relations in the UK, as city was divided from country thus enabling transgressive processes to come into being. However, while this finding might lead some to think that efforts to establish spatial divisions should now be abandoned, it was concluded that some combination of division and relation is required if nature is to be sustained into the future.

FURTHER READING

On early preservationist efforts to divide city and country see David Matless's (1998) book, *Landscape and Englishness*. On the politics of the same, see Philip Sutton's (2004) *Nature, Environment and Society*. For an alternative account set in the United States, see Adam Rome's (2001) *The Bulldozer in the Countryside*. On resource flows in and out of urban areas Herbert Girardet's (1993) book, *The Gaia Atlas of World Cities*, remains useful.

Notes

1. Sarah Whatmore, in a discussion of genetic property rights, discerns the same two environmental viewpoints: 'The one conjures a world that is hybrid "all the way down", enfolding humanity in its ceaseless commotion time out of mind. The other conjures a world until recently unmarked by the (invariably negative) "impacts" of human society, only countenancing hybridity as a technical accomplishment associated with the advent of "genetic resources"' (2002: 92).
2. While in many national contexts the equation of the countryside with nature is problematic, in England the two are seen as closely aligned. As Crandell (1993: 16) puts it: 'when we think of nature we too often conjure up images borrowed from eighteenth century England'. These images have retained their power not just through time but through space: thus, perceptions of nature in the United States and elsewhere are strongly configured by these Romantic English notions

(Neumann, 1998). It hardly needs saying that such notions are still strongly present in England itself and this justifies our equating nature and countryside in what follows (see Macnaghten and Urry, 1998, on the relationship between the two in the UK context). In general terms, this equation bears out Soper's observation that

> nature itself only begins to figure as a positive and redemptive power, and to be valued in its sublime and untamed aspects, at the point where human mastery over its forces is extensive enough for aesthetic exaltation in wilderness to replace blind animal terror. The romanticisation of nature is in this sense a manifestation of the same human powers over nature whose destructive effects it laments. (2000: 20)

3. As Sutton (2000) notes, the number of commuters into London rose from 800,000 in 1881 to 1,112,000 in 1891, while the suburban population grew from around 940,000 in 1881 to over 2,000,000 by the turn of the century. In short, the city began to reach further and further into the countryside. This bolstered the fears of environmentalists that such a sprawling of urban life would destroy the repositories of nature still lying beyond the reach of industry.

4. In the words of two experienced commentators: 'The Town and Country Planning Act, 1947, might just as well have been called the Town *versus* Country Planning Act: towns and cities were separate from the countryside and good planning would keep them so' (Cherry and Rogers, 1996: 62).

5. Membership of these groups rose on some estimates from 20,000 to over 300,000 during this period (Lowe and Goyder, 1983).

6. Greenpeace, which was only established in the early 1970s, saw its membership rise to 10,000 members in 1980 and to 400,000 by the early 1990s, while Friends of the Earth increased its membership from 2000 in 1971 to 180,000 by 1990. It is estimated that by 1981, national environmental groups in the UK had a combined membership of 1.8 million, rising to 4.2 million in 1998 (Rawcliffe, 1998: 3).

7. The link between preservationism and contemporary environmentalism is given further substance by Rome (2001), who, in an analysis of post-war suburban development in the US, argues that the issue acted as a 'bridge' between old and new environmental groups. As suburban sprawl extended out from the cities during the late 1940s and early 1950s, conservationists, environmentalists, ecologists and others began to mount local and national campaigns against destructive development. According to Rome,

> the effort also had a significant impact on the emerging environmental movement. The open-space issue pointed conservationists toward a broader more 'environmental' agenda. It created a new group of activists and a new set of grassroots organisations. Perhaps most important, the open-space issue contributed to the development of a distinctly environmentalist rhetoric and imagery. (2001: 139)

Rome (2001: 151) argues that by the 1960s the movement concerned with preserving open and green space had become closely intertwined with the movement for conserving the more general environment: 'The emergence of a popular ecological consciousness strengthened the conservation argument for open space. At the same time, the campaign for open space increased the range of support for the environmental cause'. In the process, new and old environmental groups began to develop shared interests and shared understandings of environmental problems.

8. As CPRE member, and author of an influential wartime report on rural policy, Lord Justice Scott put it in 1942: 'farmers and foresters are unconsciously the nation's landscape gardeners [...] even were there no economic, social or strategic reasons for the maintenance of agriculture, the cheapest way, indeed the only way of preserving the countryside in anything like its traditional form would be to farm it' (quoted in Green, 2002: 192).

9. Between 1961 and 1971, the population of rural areas increased by 6%, the first such net increase since the onset of the Industrial Revolution. In the following decades, the movement of population accelerated, with rural areas increasing their population share by over 9% between 1971 and 1981, and by 6% between 1981 and 1991, with even remote rural locations experiencing population increases. Between 1960 and 1987, the number of manufacturing jobs in England fell by 37.5% but the number in rural locations rose by 19.7%. This was followed by an increase in service jobs in rural locations: private service employment grew by 49% in the towns and rural areas between 1981 and 1996, but only by 19% in the conurbations (for a summary of these trends see Murdoch et al., 2003).

10. CPRE grew from 15,000 members in the 1960s, to around 20,000 members in the mid-1970s, to almost 40,000 in the late 1990s (Murdoch and Lowe, 2003).

11. Reader (2004: 297) points out that London comprises 65,000 woodlands, covering 7000 hectares, two-thirds of which is ancient woodland. As a result, in 2002, the UK Forestry Commission appointed the city's first Forestry Conservator.

12. For instance, the Commons Preservation Society claimed to have preserved 95,000 acres of common land in its first 20 years of activity (Sutton, 2004).

13. As the Environment Agency (2002) points out, many rare or threatened habitats can be found in or near urban locations: for instance, urban areas in England hold around two-thirds of the country's Local Nature Reserves, as well as large numbers of Sites of Special Scientific Interest (the most protected of ecological sites).

6

Dis/Ordering space II: the case of planning

Through exclusively social contracts, we have abandoned the bond that connects us to the world, the one that binds the time passing and flowing to the weather outside, the bond that relates the social sciences to the sciences of the universe, history to geography, law to nature, politics to physics, the bond that allows our language to communicate with mute, passive, obscure things – things that, because of our excesses, are recovering voice, presence, activity, light. We can no longer neglect this bond. (Serres, 1995)

Introduction

In the last chapter, we gained some insight into complex interactions between spatial relations and territorial zones. We saw that efforts to separate out nature from society led to the generation of transgressive and heterogeneous processes that ultimately undermined the spatial demarcations that had been put in place. It was therefore concluded that the demarcations need somehow to be aligned with relations, notably in the context of emergent ecological formations. In short, an ecological approach requires the territorialization of multiple processes, even those that extend over considerable ('global') distances.

One key means of demarcating territory is planning and, indeed, in the previous chapter the planning system came to be seen by preservationists as the primary mechanism for 'ordering' nature and society in spatial terms. Yet, despite this reliance on planning by preservationist environmental groups, planning – as a 'network of knowledge' – has traditionally had some difficulty in drawing both natural and social entities into its sphere of operation, as the last chapter also indicated. In part difficulties arise because the technological 'ways of seeing' utilized by planning tend to draw actors and entities only selectively into its governmental framework. However, planning is not only a technology of spatial management it is a political arena also. Planning decisions are made on the basis of political calculation and this too can result in very partial assessments of space being made. The upshot is that planning has considerable difficulty in 'representing' the complex and heterogeneous spaces in which it is inevitably immersed.

In this chapter, then, we examine planning in some detail in order to illustrate how planning networks generate inclusions and exclusions.[1] These inclusions and exclusions illustrate some of the theoretical points raised in Chapter 4 above, notably the differing interactions between networks and their contexts of operation. It is suggested here that planning has always been open to some entities in its surrounding environment but closed to others. Over time the entities to be included and the entities to be excluded have changed. Thus, early in its development planning successfully incorporated physical entities; it then began to shift its gaze to social entities; finally, it began to look more closely heterogeneous entities. While it has yet to successfully engage with these latter planning 'objects', there is now a recognition in planning circles that heterogeneous complexity needs to be apprehended in some way.

The case of planning thus illustrates how networks interact differentially with their spatial contexts. It shows how the constitution of the network determines relations between 'inside' and 'outside', between processes of network consolidation and processes of spatial extension. However, planning is also interesting in the context of the present volume because it comprises a form of spatial governmentality, that is, it constitutes a form of 'applied geography' in which differing conceptualizations of space are brought to bear in governmental interventions aimed at regulating the land development process. In other words, planning 'performs' or 'enacts' spaces of differing kinds (Law and Urry, 2004). Because geographical ideas are invoked within these 'performances' and 'enactments', it is important to consider the spatial imaginaries at work. In particular, it is worth asking whether planning performs topographical or topological conceptions of space so that we can then assess the potential impact of these differing conceptions on patterns and processes of spatial development.

In order to explore the spatial imaginaries at work in planning, the chapter is divided into three main parts. In the first we briefly review the key ideas about space that have shaped planning policy and practice. We chart an evolution from 'physicalist' (or topographical) notions of space to more social (or topological) conceptions. In the second part, we move on to consider how topological conceptions of space come to be bound into political processes. The discussion here focuses on the relationship between rationality and power. In Foucaultian fashion, it shows how power relations inevitably encompass technologies of spatial representation. It is then suggested that while planning increasingly recognizes *social* and political topologies, it fails to adequately appreciate the *heterogeneous* multiplicities of topological space, that is, it remains caught in the 'social contract' referred to by Serres in the quotation provided above. This point is illustrated by a brief discussion of planning for sustainable development, where it is shown that, notwithstanding an obvious desire to engage more substantively with the non-human realm, planning remains deeply rooted in socio-economic processes. As a result, it is unable to engage fully and wholeheartedly with complex urban ecologies. In conclusion, the chapter suggests that planning should abandon its traditional concern for only social entities and should immerse itself more fully in the heterogeneous complexity of contemporary urban life using a political-ecological approach. This approach specifies that politics must engage with 'collectives'. In these collectives humans and nonhumans co-exist in mutable and shifting relationships. Planning, if it is to play any kind of ecological role, must seek new ways of orchestrating such collectives so that complex and dynamic relations come to be sustained over time. Importantly, ecological orchestration requires key modifications in planning's relationship with space. These modifications are discussed in the concluding section of the chapter.

Planning and the technological 'taming' of space

For much of its history, land-use planning has sought to manipulate Euclidean spatial entities. This is perhaps understandable if we consider that conscious and deliberate attempts to intervene in urban development extend back as far as Ancient Greece. Cliff Hague (2002: 2), in a review of plan-making through the ages, claims that Hippodamus of Miletus was 'the first town planner in Europe'. In Hague's view, Hippodamus, who 'was creating regular grid layouts as early as 450 BC' (2002: 2), contributed some seminal technical ideas to planning, including procedures for the orientation of buildings and streets, recognition of the need for regular supplies of fresh water, and an appreciation of the importance of drainage. These ideas, Hague argues, endured through to the Renaissance and beyond, and they shaped the development of many European cities (see also Mumford, 1961). In short, the planning of urban Europe was, to a considerable degree, influenced by the 'geometry' of early Greek urban designs.

   Euclideanism evidently provides a theoretical underpinning to early forms of spatial planning. However, this ordered and orderly approach came under considerable pressure following the onset of the Industrial Revolution. During the eighteenth and nineteenth centuries, European towns and cities grew rapidly, with the consequence that economic and social processes appeared to outstrip any efforts at organization and control. Not surprisingly, calls were made in many newly industrializing countries for greater regulation of urban development, especially once diseases such as cholera began to cross over class lines to affect all urban dwellers. Despite these growing concerns, effective regulation of urban space only emerged onto the terrain of government in the latter half of the nineteenth century, as public health and other forms of welfare legislation were enacted in the UK and in other European states. This legislation initiated what Osbourne and Rose (1999: 738) call an 'urban will to government' and brought new forms of spatial ordering to the fore. In the UK, for instance, codes and standards were introduced to guide the development of buildings and streets during the 1870s. To implement these regulations, the country was divided into urban and rural districts under the jurisdiction of a local government management board (Ratcliffe, 1973). While the primary goal of government action was improved hygiene and sanitation, land-use planning gradually gained an important role in facilitating the orderly development of rapidly growing urban areas. As Stephen Ward (1994: 2) puts it: 'before it was anything else, town planning was a series of radical ideas about changing and improving the city'.

Strategies of change and improvement could only be implemented, however, once planners and other city officials felt they had an understanding of the entity to be changed and improved. In short, planning's 'will to govern' became 'inseparable from the continuous activity of generating truths about the city' (Osbourne and Rose, 1999: 739). In turn, these truths could only emerge once 'mundane techniques of gathering, organisation, classification and publication of information' had been put in place (1999: 739). An especially significant new technique for generating 'truths' about the city was the urban map. As Patrick Joyce explains in reference to the UK, from the middle years of the nineteenth century an innovative set of mapping techniques allowed cities to be visualized in new ways:

> [the Ordnance Survey] was in 1841 authorised by the Treasury to produce town plans [...] By 1892 urban Britain was mapped on a scale sufficient to show detail down to the size of a doorstep [...] These plans provided an unprecedented view of the city and its inhabitants. Perhaps a better term would be an unprecedented view *into* the city, for the model of vision here was the medical one of microscope, as well as the omniscient view of the surveyor. (2003: 52)

As Figure 6.1 demonstrates, the map served (in Latourian terms) as an 'immutable mobile', an inscription that translated space into diagrammatic form, thereby reducing spatial relations to a single sheet of paper. On this sheet of paper, the city would be made 'legible' – that is, it would become a place 'whose districts or landmarks or pathways are easily identifiable', because they had been 'easily grouped into an overall pattern' (Lynch, 1960: 5).

Yet, the process of rendering the city 'legible' necessarily gave rise to a very specific spatial order – that is, it held some things constant (notably, buildings and streets) and removed others from view (notably, the movement and fluidity of urban social interactions). Thus, in the map a very particular spatial formation began to emerge, as Joyce explains:

> In the plan, space is delineated, reduced to the clarity of the line. This sharpened line demarcates spaces, so that buildings, streets and so on are differentiated, but this is with reference to a common rhetoric concerning legibility [...] All the elements are different (one dwelling is sharply different from another, to a degree that is striking and new) but all are composed of the same medium, that of an extreme form of geometrical space. In this form a 'functional equivalence' is taken to new heights, in terms of the interchangeability of standardised units. One thing is read in terms of another in ways that become ever more emphatic as the standard of measure becomes ever more standard. (2003: 54)

In the nineteenth-century map, then, the city was 'drawn together': buildings, streets and open spaces were reduced to inscriptions and were then re-presented as

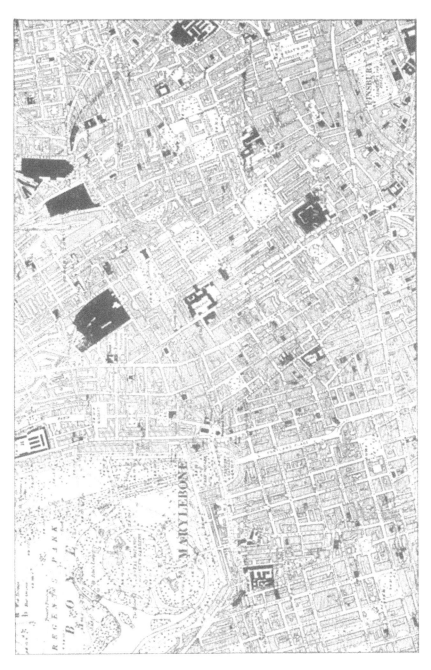

FIGURE 6.1   The gathering together of physical form in the Ordnance Survey (Source: Unwin, 1910)

lines on a page (Harley, 1988). The map comprised a novel form of connectivity, a new landscape of buildings and streets. In Deleuzian terms, mapping comprised a 'line of force' in which various entities were arranged into coherent, geometrical patterns. Moreover, these geometrical patterns were based on the physical, rather than the social, properties of cities.

Importantly, the map's geometrical and physical perspective allowed the viewer/planner to 'see' the city as a 'collective entity' (Joyce, 2003: 55). Once this entity came into being, it facilitated the practice of planning: if the city could be 'seen', then it could be moulded or shaped as political or regulatory forces were brought to bear on the visualized entity.[2] As an early planning theorist, Raymond Unwin, commented: 'In desiring powers for town planning our town communities are seeking to be able to express their needs, their life and their aspirations in the outward form of their towns, seeking, as it were, freedom to become artists of their own cities, portraying on a gigantic canvas the expression of their life' (1910: 9). The notion that planners could become 'artists of the urban realm' was a reflection of the confidence the profession gained from the new technologies of visualization: once they held the map, they held the city; and once they held the city, they felt they could mould it in line with dominant governmentalities of spatial organization.

In general terms, the refinement of these early planning technologies led to a preoccupation with topographical rather than topological space. The adoption of survey techniques provides just one example. As Raymond Unwin explains: 'The [urban] designer's first duty [...] must be to study his town, his site, the people, their requirements' (1909: 140). This would be achieved through the 'urban survey', effectively a systematic overview of all relevant urban conditions. The survey would again take the form of a map, showing, for instance, the density of population, the distribution of businesses, the location of parks, places of civic amenity, buildings of historical value, areas of poverty, areas of environmental damage and so forth. It would also attempt to model the flows of population, traffic, water and drainage in order to assess future requirements for land and other resources. Having undertaken such procedures, the planner would then be in a position to draw up a programme of future action. As another leading theorist, Patrick Abercrombie, put it: 'the survey naturally leads to the plan' (1933: 132). The plan would highlight the most important urban processes, their land-use implications and how future developments would most likely unfold. On the basis of this information, the plan could then go on to make suggestions for planning interventions – for instance, on the zoning of activities within the city, on the distribution of population in new developments and on the need to safeguard areas of special historical and environmental value.[3]

While it might be suggested that in this procedure the plan reinforced the partial spatial perspective derived from the survey, Abercrombie argued that the linking of the two would allow planning processes to unfold in accordance with 'nature'. In his view, the notion of a 'natural plan', appeals both to the descriptive and the normative sense of the word 'plan': plan understood as map, a description of the way things are; and also plan as the projection of how things should be, the prefiguration of a new state of things. The natural plan marks the coincidence of these two notions. The plan is natural in the sense that it provides an indication of the true nature of things – the way things are organised in their attachment to the natural and necessary processes by which they are determined. The natural plan is at the same time an indication of the way the environment will be once man reaches a full understanding of things and manages to act upon nature in accordance with nature. (Dehaene, 2004: 32)

In Abercrombie's work, 'the topographical map is the prime instrument used to engender a sense of naturalness in the plan, providing a seemingly neutral argument for the spatial configurations advanced by the plan' (Dehaene, 2004: 21). Yet, the use of maps not only permitted a 'natural' and 'neutral' appreciation of urban space but introduced a certain formality into understandings of spatial relations: 'the study of the city through its plan underscored an analysis of the urban morphology as a matter of formal composition, examining the formal relationships between the component parts of the plan' (Dehaene, 2004: 21). This topographical formality would ultimately deliver an 'urban syntax' or a 'structural matrix' in which natural and social processes were integrated into an organic whole.

BOX 6.1

Planning sought to 'tame' space:

- Planning required an overarching perspective on urban space, some point of view that would allow planning to envisage spatial order.
- This view was gained once urban areas were mapped, once the details of urban life were laid out in a coherent and comprehensive fashion.
- However, the map directs the attention of planners to physical features of the landscape. It thus generates a topographical, pointillist perspective that fails to apprehend the dynamic processes of 'becoming', which do so much to 'make' urban space.
- Once planning acquired some administrative or governmental authority, it could begin to act upon the spaces it could visualize but the reliance on the map meant that planning acts 'topographically' rather than 'topologically'.

Abercrombie's surveys would document 'not only the lay of the land and the composition of the underground, but also the structural features of the landscape' (Dehaene, 2004: 21). And Abercrombie was not alone in his concern for the physical characteristics of space: as Scott (1969) shows in his survey of early twentieth-century planning, planners regularly tended to gravitate towards physical attributes of cities and regions because these were more easily managed than social processes. Taylor (1998: 4) likewise explains that, until the middle years of the twentieth century, it was generally assumed that urban planning was an 'exercise in the physical planning and design of human settlements'. In his view, this 'physicalism' had three interrelated components:

1. urban planning was seen as the management of natural and physical processes;
2. design was a central aspect of such planning and this design-led approach tended to focus attention on physical form; and
3. urban planning necessarily involved the production of 'blueprints' or 'masterplans'.

Together, these three aspects of early twentieth century planning reveal a spatial imagination that focuses firmly upon the physical and objective properties of spatial formations (Abercrombie's 'natural planning').

The selective character of this spatial focus owes much to the technical resources that were being harnessed to the conduct of planning during the early phases of its development. Planning envisaged space in the form of a map. It therefore tended to demarcate and distribute activities in ways that accorded with the rationality of this mode of visualization (notably in the form of physically demarcated cells or containers). The map was complemented by the urban or regional survey, which provided a topographical overview of urban structures, processes and entities. However, the survey was again rendered into mappable form so that the physical attributes of cities and other spatial locations inevitably came to the fore. Finally, the survey heralded the plan, a summary statement on the area to be planned and the policies that would be introduced to facilitate planned outcomes. The plan served to integrate all the other technical devices into one coherent approach. It also utilized these devices in order to conjure up a technological zone (as Barry, 2001 might put it) in which governmentalities of the well-ordered space were brought to bear. Through the technologies of planning, a topographical zone comprised of physical entities was brought into being.

Planning systems

As planning gained full political powers in the post-war period, the physical zoning of space became central to governmental interventions in processes of development. However, concerns about the orientation of the new planning systems soon began to be expressed. As Taylor (1998: 14) points out, this 'tidy design conception of urban form showed no real understanding of how different residential areas actually functioned or […] of how different areas tended to develop differing patterns and concentrations of urban functions'. The very idea that 'neighbourhoods' existed or functioned as distinct entities 'was itself a design idea which had not been subjected to critical examination and, when it was [...] it was found to be "suspect"'. Taylor thus concludes that 'physicalist' planning worked simply on a series of *assumptions* about the best way to regulate complex social processes: 'the idea that the complex teeming metropolis itself might be a desirable living environment did not really come into the picture' (1998: 36).

Processes of spatial emergence, however, could not be kept completely at bay and a series of critics emerged who complained about the formality of urban planning during its formative years. Perhaps the best known was Jane Jacobs (1961), who, in her seminal study of urban contemporary life, delivered a penetrating critique of the physicalist approach. In her view, planning theory had failed to apprehend the complex and unpredictable interactions that occur between differing activities *across* urban space. In particular, planning had become disconnected from urban space:

> Cities are an immense laboratory of trial and error, failure and success, in city building and city design. This is the laboratory in which city planning should have been learning, forming and testing its theories. Instead the practitioners and teachers of this discipline (if such it can be called) have ignored the study of success and failure in real life, have been incurious about the reasons for unexpected success, and are guided instead by principles derived from the behaviour and appearance of towns, suburbs, tuberculosis, sanatoria, fairs and imaginary dream cities – from anything but cities themselves. (Jacobs, 1961: 16).

Jacobs suggested that if planning theorists were to look more closely at the city, they would see 'an intricate and close-grained diversity of uses that give each other constant mutual support, both economically and socially' (1961: 24). Thus,

> when city designers and planners try to find a design device that will express, in clear and easy fashion, the 'skeleton' of city structure [...] they are on fundamentally the wrong track. A city is not put together like a mammal or a steel-frame building, or even like a honeycomb or a coral. A city's very structure consists of a mixture of uses, and we get closest to its structural secrets when we deal with the conditions that generate diversity. (1961: 390)

In order to apprehend urban diversity, Jacobs proposed that planning theory should learn from the life sciences, in particular from the concern for 'organized complexity' in dynamic systems. In Jacobs's view this concern has three main aspects:

1. a need to analyse processes and their catalysts;
2. a requirement to work inductively, 'reasoning from particulars to the general rather than the reverse' (1961: 454); and
3. a need to look for 'unaverage clues' which reveal the way larger and more average processes are operating' (1961: 454).

Using these three interrelated ways of seeing cities, the complex but functional ordering of life in cities would be revealed.

BOX 6.2

The new interest in urban systems led to:

- A concern for the dynamic and emergent qualities of cities. These qualities were revealed most clearly by Jane Jacobs (1961). She showed the way social processes unfold across urban space in unpredictable ways. In her view, such processes were simply being ignored by mainstream planning which still focused on the physical and formal properties of cities.
- An interest in modelling urban systems using the new computing technologies that were then becoming available. These technologies held out the hope that the complexities identified by Jacobs might be incorporated into more formal planning models.
- Increasingly formalized planning models that used a great deal of quasi-scientific data to reveal the nature of urban process but still failed to apprehend socio-economic diversity in the city.
- A form of planning that seemed to hold strong continuities with early forms of technological and physical planning. Thus, the social complexities revealed by Jacobs still remained beyond planning's reach.

The *Death and Life of Great American Cities* (Jacobs, 1961) provided a sustained critique of the 'physicalist' planning practiced by Unwin, Abercrombie and others. But more than this, it opened up a new means of understanding cities based on an appreciation of complexity and diversity. This latter insight was subsequently adopted by a new generation of planning theorists who had become interested in systems theory (McLoughlin, 1969). Like Jacobs, the systems analysts believed that cities should no longer be seen as simple zones on a map but should instead be viewed as complex clusters of interacting processes (Chadwick, 1971). While the systems approach effectively built upon the earlier idea of the urban survey, it specified that the various economic, social and environmental processes incorporated into any urban overview should be formally assessed in systemic models. It thus adopted a seemingly dynamic perspective on urban process.

The models, by demarcating new urban zones in which various entities were brought into complex and dynamic relationships, apparently promised an advance on the topographical approaches used during the early years of planning. In

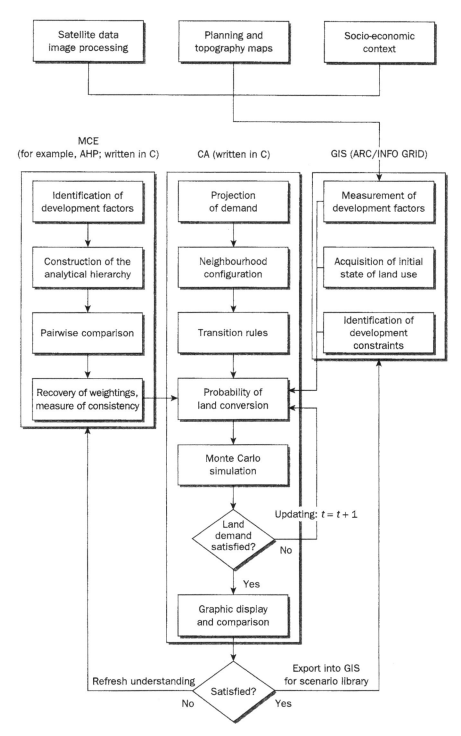

FIGURE 6.2 A contemporary example of systems theory applied to planning (Source: Wu and Webster, 1998)

particular, the new technology of computer modelling seemed to hold out the hope that the city could be represented as a complex set of dynamic and interconnected processes. However, as compuional methods were developed they tended to become formalized so that 'bulky plans, previously spruced up with photographs, sketches, designs and land-use maps, became even bulkier 'scientific' studies replete with statistical and mathematical appendices and technocratic jargon' (Rodwin, 2000: 16). The increasingly technical character of the models generated some amount of unease amongst planning practitioners. There were two main sets of concerns. First, despite the computing power then becoming available, the models still seemed simplistic; that is, they were only able to capture a limited number of urban processes, in part because key data sets associated with economic and social entities in the city were still unobtainable (for instance, data on incomes, rents and ownership were notoriously difficult to get hold of but were crucial to the functioning of the models) (see Batty, 1985). Thus, the predictions derived from the models proved inherently unreliable. Second, the models failed to adequately account not only for the socio-economic aspects of urban development but also for the likely impact of any planning interventions on these aspects. In short, computers, like the earlier technical devices used in the planning process, tended to create 'technological zones' in which entities were arranged not in ways that reflected topological complexity but in ways that reflected the topographical simplicities enshrined in the planning models themselves (Faludi, 1973). In other words, while the new technologies allowed more entities to be drawn into the planner's line of sight, they still tended to produce physical and geometrical spatial perspectives.

Planning the social

The formalistic character of the new computer-aided models was of particular concern to those theorists who concerned themselves with planning as a social and political process (see, for instance, Breheny and Hooper, 1985). These new theorists argued that planning had become increasingly politicized following the establishment of state planning agencies in the post-war period. Thus, while planners could use rational and sophisticated technologies in order to initiate appropriate development patterns, they were ultimately immersed in some form of political bargaining (Friedmann, 1987). And as planning settled into its governmental role, any hard and fast distinction between planning as a technical activity and planning as a political activity began to dissolve. Charles Hoch summarizes the implications of this:

> Professional planners face a serious problem in our liberal democratic society. Their professional judgement relies on theoretical and specialised knowledge of social, economic, political and geographic relationships. However, most of the problems that planners analyse and assess are practical. Officials, advocates, lobbyists, and citizens possess

specific attachments that make them feel as if they are experts on the practical matter at hand [...] Many planners understand this dilemma and work hard to solicit and include in their professional advice the ideas and opinions of the citizens involved. However, the pursuit of widespread consensus through compromise and negotiation often seems to undermine the integrity of expert judgement. If people can figure out what they want and how to get it on their own, who needs professional planners? (1994: 1)

Thus, once planning achieves legislative status, planners must link their specialized knowledges to the demands of politicians, developers, environmental and amenity groups and other planning participants. In the view of many planning theorists, this linking cannot be achieved using formal and rational models; rather, it requires a reflexive approach that somehow incorporates both technical and political perspectives within a new form of social engagement. The American planning theorist, John Forester, identifies the importance of such engagement when he says:

In cities and regions, neighbourhoods and towns, planners typically have to shuttle back and forth between public agency staff and privately interested parties, between neighbourhood and corporate representatives, between elected officials and civil service bureaucrats. They do not just shuttle back and forth though. Trying to listen carefully and argue persuasively they do much more. They work to encourage practical public deliberation – public listening, learning and beginning to to act on innovative agreements too – as they move project and policy proposals forward to viable implementation or decisive rejection. (1999: 3)

As Forester indicates, the recognition that planners are immersed in complex political sub-fields shifts the emphasis away from the achievement of rational outcomes towards the management of processes of decision-making. The planner is no longer seen as the neutral facilitator of expert interventions but as the orchestrator of political processes, processes that comprise multiple actors and multiple viewpoints. As Hoch (1994: 105) puts it: 'planners do not uncover facts like geologists do, but rather, like lawyers, they organise facts as evidence within different arguments [...] all engage in persuasive rational arguments [...] focused and attached to value objectives'. In other words, planners do not stand outside political processes but play a full part as interested participants. Thus,

the planner could serve not only as a designer and coordinator but as advocate, negotiator, or coalition builder. It was equally respectable to characterise the planner as knowledgeable not only about the problems of urban land use and environmental policy but as someone with generic skills in policy and analysis, the processes of communication and negotiation, as well as implementation and public management. (Rodwin, 2000: 16)

In thinking through the implications of the planner's role as a coalition builder or negotiator, planning theorists began to turn to alternative technologies of representation. In particular, they began to use qualitative approaches such as ethnography and participant observation, which had been pioneered in sociological analyses of urban life. In the 1920s and 1930s, the so-called 'Chicago School' of urban sociology had established the importance of qualitative research for the study of social groupings while, during the same period, anthropologists such as Boas, Mead and Malinowski had shown how detailed ethnographic studies might be employed to understand 'alien' cultures (see Denzin and Lincoln, 1994). All these qualitative approaches delved into the lifeworlds of participants and brought multiple social perspectives to the surface. They allowed social complexity (for example, the differences *between* urban lifeworlds) to be more easily appreciated than had been the case in the formal models and surveys. In short, their use brought the 'social' into view and rendered it amenable to processes of planning governance.

The sociological perspective was 'governmentalized' through 'communicative' or 'collaborative' planning. The basic idea here is that planners should aspire to giving all participants ('stakeholders') a voice in planning processes (Bryson and Crosby, 1992). According to Healey (1998: 312), a leading exponent of the collaborative approach, 'those involved as experts in [planning] processes should have an ethical duty to attend to all stakeholders as the interactive process develops'. In order to draw stakeholders into decision-making procedures, planners need to engage in such activities as:

1. listening hard to a variety of participants;
2. cultivating community networks;
3. finding ways of involving uninterested but affected social groups;
4. educating citizens about planning choices; and
5. encouraging community groups to develop their own planning proposals (see Forester, 1989, 1999).

The reform of planning processes along communicative or collaborative lines also requires the consolidation of an institutional structure that helps to build the potential for genuine participation. Such a structure would assist collaboration,

> by its role in informing political communities about the range of stakeholders and about how they would like to discuss issues; by helping to shape arenas where stakeholders can meet; and by helping those involved work out what it means to build new collective ways of thinking and acting, to reframe and re-structure their ways of proceeding. (Healey, 1998: 312)

A key part of such institutional reform involves the making of plans. No longer can plans be seen as detached overviews ('*master*plans') of sharply delineated spaces; rather they become 'processes of interaction' between participants (Healey, 1993: 83). Plan preparation itself is renewed as a 'process of "making story lines"' so that all participants should be permitted to see their interests reflected in the final version of the plan. As Healey (1993: 103) puts it, the land-use plan becomes 'an expression and record of the strategic choices and moral dilemmas a community has to face. This means consideration not only of the struggles between the community and other communities, but between the diverse cultures and systems of meaning within a community' (1993: 103). This amounts to a 'socialization' of planning's technologies.

BOX 6.3

Graham and Healey (1999) propose that communicative and collaborative planning might be thought of as 'relational planning'. The approach has four main components:

- First, planning should consider relations and processes rather than objects and forms. In practice this means that 'the extent to which a proposed form [...] will lead to particular social, economic, and cultural behaviours needs to be demonstrated in terms of the relational dynamics of specific instances, not assumed as a universal generalisation' (1999: 642).
- Second, planning should stress the multiple meanings of space and place associated with differing social groups and differing social identities. This requires 'careful assessment of the many spatial and temporal experiences of a city, and how these flow across and into each other in shaping a place and filling it with value' (1999: 642).
- Third, planning needs to consider specific spaces as 'layers of relations'. In identifying such layers, planners should recognize that privileging one experience of space and time may necessarily undermine others.
- Finally, planning should promote communication between differing social groups and networks in order to reduce social exclusion. Planners thus need to facilitate the recognition of these and should also help mediate inevitable conflicts 'without allowing one-dimensional viewpoints to regain their dominance' (1999: 642–3).

The emergence of communicative and collaborative discourses indicates that planning is moving away from a prescriptive and rational ordering of space (using technical models and plans) towards a more social process of decision-making based on understandings of cultural, political and ethical differences. This move implies a concern not for the smooth and uniform spaces of the map, but an interest in the undulating and varied spaces associated with social entities as they move through the city. Thus, communicative and collaborative theorists work to produce new planning technologies that encompass diverse social entities. As Leonie Sandercock (2003: 76) observes, the development of such technologies entails an engagement with an 'epistemology of multiplicity', one that recognizes various ways of 'knowing' space (listening, dialogue, contemplation and so forth). Sandercock argues that this epistemology requires 'new models of planning practice which expand the language of planning beyond the realm of instrumental rationality' (2003: 76). In her view, a transformed language would usher in a new spatial imagination so that plans and other planning interventions would be based on an engagement with the spaces of 'lived experience' rather than the (dead) spaces of maps and models. Planners would then be open to the 'emotional breadth and depth of cities [...] cities of desire, cities of memory, cities of play and celebration, cities of fear and paranoia, cities of struggle' (2003: 227). As space becomes socialized, a zone of multiplicity and relationality inevitably comes into view.

Planning politics

The above discussion indicates that planning theory moved some considerable distance over the course of the twentieth century, from the physical planning of Euclidean or topographical space to the planning of social or topological diversity. This move was driven, in part, by a normative concern to build up the scope for social inclusion and consensus generation so that planning processes could represent the full range of urban stakeholders. This culminates in an 'epistemology of multiplicity' in which differences in the life-worlds of urban inhabitants are recognized by planners in both their plan-making activities and in the decisions they make on specific developments. Planning therefore becomes a process of 'orchestration', in which the planner endeavours to encourage both an open dialogue and the achievement of some form of consensus between involved stakeholders. Yet, while the assertion of this socially reflexive form of planning has generally met with a great deal of support from planning theorists and practitioners, it has also aroused the suspicions of those who see planning as inevitably dominated by politics and power (for example, Richardson, 1996; Tewdwr-Jones and Allmendinger, 1998; Flyvbjerg and Richardson, 2002). This latter group is sceptical of any claim that planning can somehow 'exceed' the powers that 'produce' planning decisions (as Foucaultian theorists might put it). Thus, any move to construct a planning consensus needs to be taken in the full knowledge that it may simply involve yet another imposition of power relations, culminating in new spaces of domination and subjection.

The leading exponent of this 'political' perspective is Bent Flyvbjerg. In a detailed and nuanced study of 'planning in action', Flyvbjerg (1998) shows that planning is a political activity through and through. In particular, he argues that any (technical) rationality asserted by planners is *always* set within relations of power ('rationality is context-dependent and [...] the context of rationality is power', 1998: 2). In outlining this broadly Foucaultian perspective, Flyvbjerg presents a detailed case study set in Aalborg, a medium-sized city located in the North Jutland region of Denmark. In what follows, we will briefly review the main findings from Flyvbjerg's study, as these help to illustrate the uneasy relationship between planning's technical and political functions. They also indicate a need for planning to recast itself as an 'immanent' rather than normative process of spatial regulation.

Flyvbjerg begins his evaluation of planning in Aalborg by examining the arguments that unfold around the siting of a new bus terminal as part of a reorganization of the city's transportation system. He notices that as the decision on the location of the terminal begins to take shape, political rather than technical considerations come to the fore (it seems the bus company favours one particular site and uses its political influence in the project task force to ensure this site is chosen, irrespective of its technical merits). Once a decision has been reached, however, some justification is required. Flyvbjerg summarizes the sequence of events thus:

> it transpires that even before the technical evaluations of placement options for the terminal have been completed, the Aalborg Project's Task Force decides to locate the terminal at Nytorv [...] What happens, then, is that the decision regarding the location of the bus terminal is made simultaneously with a decision about elaborating the technical basis for the decision. (1998: 21).

In short, technical expertise is employed to 'rationalize' a decision that has already been taken on political grounds. Flyvbjerg is therefore able to suggest that 'the rationality produced is actively formed by the power relations which are themselves grounded and expressed in processes that are social-structural, conjunctural, organisational and actor-related. Conversely, these power relations are supported by the rationality generated' (1998: 27).

The decision on the siting of Aalborg bus terminal is only the first in a whole series of negotiations around the city's transportation system that are documented by Flyvbjerg. In these negotiations, the main aim of the planning authority is to achieve some reduction in the impact of automobiles on the city's environment. In order to restrict the number of cars entering Aalborg, the planners formulate a 'traffic zoning solution' – that is, they divide the central area of the city into four discrete areas, which are all separated by barriers (to be reached via a ring road). In each area, a balance is created between the supply and demand for parking, with the streets made accessible to other modes of transport, notably buses, bikes, pedestrians. The planners argue that this zoning of traffic will lead to fewer injuries and deaths, less traffic noise, and reduced air pollution.

The planners thus use a form of spatial zoning in order to control key socio-technical processes such as the flow of traffic through the city. In this regard, they strive to align topographical and topological perspectives – that is, areas can be zoned so that the flow of cars is effectively regulated. However, they believe that this alignment will only be successfully achieved if the plan is implemented in its entirety, all at once; they worry that if the negotiations become too protracted the scheme will begin to fall apart. Their concerns seem justified for politics soon enters the picture again in the form of the Aalborg Chamber of Commerce. The Chamber is worried that restrictions on traffic may affect retail sales. It therefore opposes the introduction of the traffic zones and asserts the need to look in more detail at the impacts of the proposals on retailing. According to Flyvbjerg, the staff in the Technical Department of the City Council believe that the Chamber's opposition to the project is based on very weak (technical) arguments (for instance, the Chamber simply has no evidence to back up its claims about the negative impacts of the zoning scheme). Yet, the city's technical staff is not allowed to argue with the Chamber over the sub-stantive details of the latter's objections; rather, a negotiation is conducted between leading members of the business community and key politicians. Following this negotiation, the scheme is amended: now the implementation process will not be 'once and for all' (it will take place in stages) and the zoning scheme will only be partially installed. Flyvbjerg suggests that at this juncture the project begins to fall apart.

> the functional coherence of the project becomes [...] more and more neglected. An integrated plan becomes a fragmented reality. In more general terms, the problem can be expressed like this: that which ought to be a rationality-to-power relation, if the rel-evant technical functional linkages in the project are to be ensured, instead becomes a power-to-power relation, where functional considerations become subordinated to tactical ones. (1998: 81)

However, even this diminished scheme meets further political opposition once it gets close to implementation. Again, the Chamber argues for the lifting of almost all restrictions on traffic. After yet further rounds of negotiation, the project finally reaches the City Council for ratification. But by now, Flyvbjerg argues, years of obstruction and argument have left it in a fragile and dimin-ished state – the original plans for a traffic-free Aalborg have given way to a set of disjointed and denuded proposals that will seemingly fail to make any sig-nificant impact on the amount of traffic in the city centre (during the period in question, the amount of traffic in Aalborg rises by 8 per cent). In other words, the simplicities of the zoning system have largely been abandoned in favour of uninhibited (traffic) flows.

Flyvbjerg describes the Aalborg Project in detail because he believes it shows how planning, as a rational and technical activity, is deeply implicated in power relations. In this sense he follows in Foucault's footsteps. He claims the Aalborg study reveals 'rationality to be a discourse of power [...] Rationality is pene-trated by power, and it becomes meaningless, or misleading – for politicians, administrators and researchers alike – to operate with a concept of rationality in which power is absent' (1998: 227). Moreover, rationality is enmeshed within relatively stable and enduring power relations. For instance, Flyvbjerg notices that the influence of the Chamber of Commerce over city planning is not unique to the Aalborg project but is a routine aspect of the planning process. As he says, 'through decades and centuries of careful maintenance, cultivation and reproduction of power relations, business created a semi-institutionalised posi-tion for itself with more aptitude to influence governmental rationality than was found with democratically elected bodies of government' (1998: 227).

BOX 6.4

Flyvbjerg (1998) proposes a political perspective on planning processes. This perspective shows:

- That planning is saturated with power relations, in much the way that Foucault would lead us to believe.
- These power relations can be seen as two main types: straightforward political power and power that derives from technical rationalities.
- In Flyvbjerg's study, political power constantly seems to have the upper hand against technical forms of power.
- Political power is exercised by local politicians and local business owners; technical power is exercised by the planners.
- Flyvbjerg's findings mean that we must pay particular attention to (political) power relations when assessing scope for collaborative or communicative planning.

In the main, Flyvbjerg is concerned to show that the assertion of any normative agenda within land-use planning must engage with the reality of planning's power – that is, with the routine assertion of power relations in planning processes. So while the communicative or collaborative planning theorists may wish to open up space for multiplicity in the making of planning decisions, the likelihood is that such efforts will routinely encounter strategies of exclusion and manipulation. Flyvbjerg therefore suggests that instead of seeing revised planning processes as a means of dissolving power, 'we need to see them as practical attempts at regulating power and domination' (1998: 236). In other words, the best that can be achieved is the enhancement of alternative planning powers, powers that give voice to currently marginalized groupings. We see here, then, that planning opens up a political space, one that is striated by power relations of various kinds.

Flyvbjerg's study clearly shows how difficult it is to separate the technical and political dimensions of planning – the two become inextricably bound into one another. In short, it indicates that planning is deeply embedded in a fully politicized space. It also infers that planning must be seen not as some normative ideal (as in the communicative and collaborative planning approaches) but as 'immanently' ensmeshed (as in Foucaultian and Deleuzian approaches) in the socio-political processes it seeks to challenge. Flyvbjerg seems to believe that a recognition of planning's 'immanence' will allow both planning theorists and planning practitioners to more easily appreciate the need for planning norms that somehow incorporate modes of critique (for instance, of relations of power).[4] However, in the latter stages of Flyvbjerg's study another form of immanence comes into view. Flyvbjerg describes it thus:

> The losers in the struggle over the Aalborg Project are those citizens who live, work, walk, ride their bikes, drive their cars, and use public transportation in downtown Aalborg, that is, virtually all of the city's and the region's half million inhabitants. Every single day residents and communities in downtown Aalborg are exposed to increased risk of traffic accidents, higher levels of noise and air pollution, and deteriorating physical and social environment. (1998: 223)

We begin to get a glimpse here of what might be termed the 'Aalborg assemblage', the complex mix of entities that ultimately form part of the planning and political networks that Flyvbjerg so assiduously documents in the offices and meetings of the city council and the Chamber of Commerce. Rationality may ultimately be enmeshed in politics, but politics is ultimately enmeshed in materialities of various kinds. So we can suggest that the assertion of rationality and power in Aalborg has real material impacts in the city itself: the outcomes ultimately take shape on the ground, in the form of streets cluttered with cars, high levels of noise, decreased sociability, rising levels of air pollution and increased consumerism. The Aalborg transportation project as described by Flyvbjerg defines a very particular type of urban space: not a space of open, inclusive multiplicity but a space of narrow closed self-interest in which economic definitions of spatial utility (for example, retailing) win out over all others to the detriment of the wider Aalborg environment.

Environmental planning

Flyvbjerg's study shows how planning can remain relatively detached from the material environment: it exists mainly in the world of political and economic calculation. It also reveals how planning decisions affect (or fail to affect) complex urban assemblages. It might therefore be assumed that while planning incorporates spatial imaginaries (as governmentalities) it fails to incorporate space itself (as a complex set of interacting entities). Yet, this assumption would seem a little premature, for in recent years considerable effort has been expended in embedding planning more fully in heterogeneous environments. Perhaps the most successful of these efforts has been 'planning for sustainability', a mode of planning that can be traced back to the World Commission on Environment and Development, chaired by Gro Harlem Brundtland. The Commission produced a report in 1987 entitled *Our Common Future*, which identified the urgent nature of the environmental problems facing the world. The solution to these problems was proposed as 'sustainable development', a notion that was defined in rather abstract terms as 'development that meets the needs of the present without compromising the ability of future generations to meet their own needs' (WCED, 1987: 43). All decision-makers in all tiers of government were urged to take this principle on board. As Luke (1999: 142) puts it (after Foucault), sustainable development 'engenders its own forms of "environmentality", which would embed alternative rationalities beyond those of pure market calculation in the policing of ecological spaces'. This new form of 'environmentality' would focus on 'establishing the right disposition of things between humans and the environment' (1999: 146). Planners were thought to have a key role to play in enforcing this disposition: as Susan Owens (1994: 440) puts it, 'planning and sustainability share two fundamental perspectives – the temporal and the spatial. Both are concerned with future impacts on and of particular localities'. In short, planning for sustainability can not only conserve land and other natural assets for future generations, but can also ensure that any development that does take place will have a diminished impact on the overall environment (for instance, in terms of pollution and waste) (Blowers, 1993). Planning is thus a key form of 'green' governmentality.

In an era of sustainable development, planning is forced to confront more than just socio-economic relations – it must somehow encompass the 'wider environment', including 'ecological' resources and processes. Owens and Cowell (2002), in a detailed study of planning for sustainability in the UK, consider how planning authorities have attempted to bring these resources into their decision-making processes. The authors identify a number of techniques – environmental appraisal, cost-benefit analysis, environmental forums – that have been employed to 'give voice' to natural entities. They suggest that perhaps the most effective technique for drawing environmental factors into planning decisions is the environmental capacity study. The concept of environmental capacity effectively rests on the idea that 'there are some absolute constraints to development in a locality, beyond which one cannot go without unacceptable change occurring' (Rydin, 1998: 749–50). As Rydin notes, this notion draws heavily on 'the concept of carrying capacity as used in ecology and biological sciences, in which a relationship is posited between the ability of an ecosystem to support a species and the size of the species population' (1998: 750).

For planning to adopt such an ecologically inflected approach would seem to indicate that there is now a firm commitment to linking planning processes more fully to the heterogeneous materiality of space. However, Rydin suggests caution in any such interpretation, for,

> it is clearly not the case that a locality in Britain has a finite capacity for supporting a human population in the same sense that an area of open land has a capacity of a given population of rabbits, say. Extremely high densities are possible for human populations; technological investment, evolving modes of social organisation and the ability to import many of the necessary resources and export much of the resulting waste all render urban development potentially unaffected by local capacity constraints. (1998: 750)

Given the absence of any *absolute* limits or capacities, decisions need to be made about the spatial entities that should be protected in line with principles of sustainable development – 'hence the emphasis on identifying those elements of the local environment which future generations should be entitled to enjoy and using the land use planning system to protect them' (1998: 750). But making such decisions is likely to be fraught with problems. Owens cites two main sets of difficulties:

> First in the politics of development nothing is sacrosanct whatever might be theorised about the need for stringent protection of critical environmental capital. A second difficulty is that the delineation of environmental capacity will never (and should never) be purely an exercise in technical rationality. Deciding what is sufficiently important to be left intact for future generations or for its own sake, or what aspects of environmental quality should be maintained, involves judgement and makes the resultant interpretations of sustainable development particularly vulnerable to critique. (1997: 296)

Thus, we are back once again in socio-political decision-making in which environmental entities are either protected from, or sacrificed to, development. Although some commentators have sought to set limits to any trading away of the environment – by, for instance, suggesting that 'critical natural stocks' be excluded from any bargaining process (see Jacobs, 1997) – in practice, capacity studies, like all the other techniques of planning for sustainability, ultimately take their place in the 'to and fro' of planning politics. Owens and Cowell conclude that,

> to conceive of these approaches as instruments in promoting some preformed, consensual concept of sustainability is profoundly misleading. In practice, all are bound into power struggles in which conceptions of what is sustainable are actively constructed and negotiated. And since these struggles are invariably unequal, no assessment of the role of appraisal, or participations, or integration, is likely to be adequate if it is divorced from fundamental questions of agency and leverage over the political process […] what emerges as 'sustainable' depends on the arrangement of actors, opportunities and constraints in any given setting. (2002: 70)

We return here, then, to the Foucaultian assumption that planning is also an exercise of governmental power and that those able to stabilize the most coherent and robust sets of power relations are those that are most likely to prevail in defining the significance of environmental capacity and other such environmental designations.[5]

BOX 6.5

The introduction of sustainable development policies has led to the emergence of environmental planning:

- This form of planning works to draw natural entities into planning processes, so planning consists not just of calculations concerned with economic, social and political criteria but with environmental or ecological criteria as well.

Continued

- Planning has developed a number of mechanisms to draw environmental entities into the process of calculation: environmental appraisal, cost-benefit analysis, capacity studies and so forth.
- However, it still remains the case that planning privileges economic, social and political criteria. Thus planning fails to engage with the heterogeneous materiality of space. It also fails the recognize the processes of 'becoming' that emerge from heterogeneous materialities.
- Thus, planning needs to become more 'ecological in focus'; it needs to find ways to bring natural and social phenomena within manageable collectives. It needs to find ways of building robust relationships between entities of different types so all can be sustained over time and through space.

Ecologizing planning

The case of planning for sustainable development shows that although efforts are currently underway to press planning authorities into a fuller recognition of the value of environmental entities, ultimately these efforts end up becoming subject to political and social negotiation. The environment becomes a 'tradable good', simply one more factor to be weighed in the balance as decisions are made on the composition of plans and on the benefits of specific developments to differing social groupings. Planning thus continues to adhere to a 'social' rather than a 'natural' contract (Serres, 1995). Moreover, planning will only establish a 'natural contract' once it finds a better balance between physical, social, political and environmental factors in its decision-making processes (as we have seen above, planning theory seems to be able to focus on only one of these factors at a time). In short, there is a need for planning to become more fully ecological in scope.[6]

Some clues as to how planning might be made more 'ecological' are provided in Bruno Latour's (2004) book *The Politics of Nature*. In this work, Latour encourages those interested in the heterogeneous complexity of space to rethink the fundamental relationship between politics and ecology. He argues that the aim of political ecology is not to root politics in nature but rather to 'convoke a single collective' (2004: 29), made up of 'associations of humans and nonhumans', associations in which humans and nonhumans 'exchange properties' (2004: 61). All that matters, in this ecological approach 'is the production of a common world, one that [...] is offered to the rest of the collective as an occasion to unite' (2004: 141). We can see here that Latour is implicitly invoking Deleuze's notion of 'a line of force' as something that can orchestrate heterogeneous entities into an alignment that produces a common world. Importantly, Latour identifies the need for both technical (scientific) procedures and political strategies in the production of common worlds. He argues that technics and politics both hold the wherewithal to give voice to humans and nonhumans alike:

Politicians and scientists all work on the same propositions, the same chains of humans and nonhumans. All endeavour to represent them as faithfully as possible [...] both callings delight in the art of transformations, the [scientists and technicians] to obtain reliable information on the basis of the continual work of the instruments, and the [politicians] to obtain the unheard-of metamorphosis of enraged or stifled voices into a single voice. (Latour, 2004: 148)

Through the use of technical procedures, Latour believes scientists and technologists can draw nonhumans into the collective, while through the use of political procedures politicians can establish the linkages that build a single collective. He summarizes the significance of the new perspective by saying: 'it was thought that political ecology had to bring humans and nature together, whereas it actually has to bring together the scientific [technical] and political ways of intermingling humans and nonhumans' (2004: 148).

BOX 6.6

Fitzsimmons (2004) follows Latour (1999) in assessing the means by which ecology as a scientific discipline succeeds in enrolling nature into its networks:

- First, it *mobilizes* nature using field observation, methods of measurement, means of conceptual categorization and so forth. As entities are mobilized so they take their place in the world of ecologists.
- Second, it determines its own ability to speak 'for' nature through processes of *autonomization* – that is, the development of institutional structures and their associated knowledge systems.
- Third, ecology must build *alliances* with other groups and bodies in order that it may garner, resources, recognition and influence.
- Fourth, it must seek to gain *public representation* in the wider social field so that programmes for action are widely disseminated.
- We see in these four manoeuvres some of the means that planning might use to draw nature into its field of action. In particular, we see that planning needs to find more efficient and effective means of mobilizing nature as well as enhanced strategies for the construction of alliances that might sustain nature.

These various aspects of the ecological approach allow us to usefully reflect upon the resources open to planning – itself a techno-political enterprise – in building ecological spaces. As we have seen above, planning recognizes physical, social, political and environmental spaces but fails to bring these together in any kind of coherent theory or vision. Latour appears to suggest that rather than seeking to superimpose these various factors one upon the other, planners should attempt to rethink the distinctions between the physical, the social, the political and the environmental. Now they should recognize 'there are no clear boundaries, no well-defined essences, no sharp separation[s]'; there are instead 'tangled beings [...] rhizomes and networks' (Latour, 2004: 24). Thus, planning needs to draw diverse entities and processes into heterogeneous collectives in which new, more complex relations are established between all participants. In drawing these aspects together, planning needs to harness both technical and political resources: the technical to visualize and demarcate the many heterogeneous features of the spaces to be planned (perhaps in the style of Abercrombie's 'natural plan'), and the political to ensure that these heterogeneous features are brought into some kind of regulatory alignment (perhaps in the style of communicative or collaborative planning).

Moreover, in undertaking this task, planning needs to employ more than simply 'foresight'; it must also be prepared to engage with uncertainty for no one can say for sure what form these collectives will take or where the processes of collective development ('becoming') will lead. As Latour puts it:

> Political ecology does not shift attention from the human pole to the pole of nature; it shifts from *certainty* about the production of risk-free objects (with all their clear separations between things and people) to *uncertainty* about the relations whose unintended consequences threaten to disrupt all orderings, all plans, all impacts. (2004: 25)

Working with uncertainty is a new challenge for planning but it flows inexorably from the engagement with heterogeneous materialities and all their complex and unpredictable interrelations. Thus, plans need to be recast in order to forsee perhaps multiple trajectories of change, with the proviso that even this more complicated form of foresight is hedged with ambiguity and doubt.

We can see then, that the planning of heterogeneous or ecological relations requires some considerable modification in planning's *modus operandi*. It implies that a new round of innovation in planning theory and practice is required so as to straddle technical-socio-political divisions. As Latour (2004: 69, original emphasis) indicates, planning needs to 'traverse the now–dismantled border between science and politics, in order to add in a series of new voices to the discussion, voices that have been inaudible up to now [...] *the voices of nonhumans*'. The importance of this widening of planning's franchise stems from the fact that planning, as a mode of 'green' governmentality, can only be seen as truly legitimate once all the entities affected by its interventions are included in the relevant decision-making processes. Extending the franchise to nonhumans, however, requires attention to two processes simultaneously: an enlargement of planning procedures through the employment of techniques that 'give voice' to nonhumans (the environmental capacity study is perhaps an early illustration but this needs to be augmented by innovations such as ethnographic studies of human–nonhuman relations – see, for instance, Descola, 1992) and the linking of heterogeneous entities within plans in ways that ensure a stable collective comes into view (using perhaps political techniques of consensus building but now broadened to include non-human participants). If these two aspects can be sympathetically combined then planning might at last succeed in opening up a truly ecological, that is, heterogeneous spatial realm. Within this realm new heterogeneous alignments will be brought progressively into view and, once seen, they will be acted upon in ways that sustain their collective properties. The purpose of any planning action should not, therefore, be the simple ordering of (ecological) spaces but rather should comprise the nurturing of new assemblages in ways that allow orders and disorders to coexist. This requires, in turn, the development of a new spatial imagination for, as Latour (2004: 165) reminds us, we are now no longer dealing with 'closed, concentrated spaces' but rather with 'flowing basins, as multiple as rivers'. This new spatial imagination will thus need to acknowledge the 'flowing' and 'multiple' character of topological space.

Conclusion

As a form of applied geography, planning plays an important role in formulating ideas about space. Moreover, as a part of the state, planning has the opportunity to put these ideas into practice. It thus comprises a key means by which spatial imaginaries are 'performed' or 'enacted' (Law and Urry, 2004). Through its land-use plans and the decisions it makes about specific developments, planning plays a key role in ordering space. It sets out a vision of space in maps, surveys and plans, and then develops policies in order to shape trajectories of spatial development over time. The conceptualizations of space used within planning thus have important and tangible effects; they frequently appear *in space* as a result of political interventions in the development process. Planning shapes given spatial locations in line with its own views of what the 'well-ordered' zone should look like.

We have seen in this chapter some of the spatial imaginaries that have been used to conceptualize 'well-ordered space'. These extend from the Euclidean notion of spatial containers, in which socio-economic processes can be encased, through social perspectives on the multiple lifeworlds that co-exist within urban areas, to the heterogeneous materialities of environmental and ecological spaces. These varied spatial imaginaries indicate that planning holds a variable relationship *with* space. It views space through technological and political mechanisms that *select* the spatial attributes thought to be of most significance and intervenes *in* space on the basis of this selection. Planning therefore holds very partial spatial linkages and fails, in the main, to engage with the full range of entities to be found within discrete spatial locations. In Michels Serres's (1995) terms, it relates to space through a 'social' rather than a 'natural' contract.

In the previous section, we put forward some tentative suggestions for a revised planning approach, one that engages not just with the human realm but with the nonhuman realm too. By broadening out the very partial views of space that have predominated in planning circles, it has been argued that planning can immerse itself more fully in the heterogeneous materialities of ecological space. It can become *immanent* to the places and spaces being planned. In so doing, planning might re-invent itself as a form of ecological 'steering' in which multiple trajectories of development are defined simultaneously. The goal of planning then becomes sustaining the collective or the assemblage, ensuring that the rich sets of linkages that bind humans and nonhumans together are allowed to develop in ways best suited to the entities and their alignments – that is, the planner no longer 'knows best'; s/he learns from the collective what is in the best interests *of* the collective. This form of planning would no longer be seen as *master*planning – rather it would involve such activities as 'collecting', 'mixing' and 'sustaining'. Some of these activities are already undertaken by planners but the way they are conducted needs to be rethought in the light of the new ecological requirements. In short, planning must face up to heterogeneity, to the full range of entities now included in the collectives.

While these suggestions appear to chart a new course for planning – a course that simultaneously involves a new relationship with space – there are few signposts available showing how planning might reach its new destination. For instance, Latour's (2004) work on political ecology is strong on *re-conceptualizations* of science and politics in the wake of political ecology but is weak on the specific steps that might be taken to shift scientific and political *practices* in the desired direction. Thus, the means by which planning might reconfigure its relationship with space in order to engage with heterogeneous materiality remain hard to discern. While in this chapter we have identified the need for a new spatial imaginary we have been unable to suggest how this might be developed. In the next chapter, we hope to overcome this weakness by turning to examine a little more closely how new spatial relations might be brought into being not just in theory but also in practice. In so doing, we look closely at the network forms that might be required if a new interaction between conceptualizations of space and the implementation of those conceptualizations in actual, material spaces is to be achieved. The example provided, however, concerns not planning but food.

SUMMARY

This chapter has given an overview on the spatial imaginaries at work in planning. It has been argued that planning, as an arm of modern government is in a position to bring certain geographies into being – that is, planning can 'perform' space in the decisions it makes about development. Various spatial imaginaries were identified. The first were early forms of mapping the urban realm in order to show its physical characteristics. However, these in turn were replaced by more social perspectives so that planning came to focus on the management of inclusive processes. This inclusion has only recently been extended to nonhumans, however, mainly because planning remains a site for the play of (Foucaultian) power relations. In conclusion, it was suggested a new approach, derived from political ecology, should be adopted so that planning might engage more wholeheartedly with heterogeneous complexity.

FURTHER READING

Flyvbjerg's (1998) book, *Rationality and Power*, discussed at length above, remains the best treatment of power in planning. For an alternative view, from a communicative planning standpoint, see John Forester's (1989) book, *Planning in the Face of Power*. For some thoughts on relationalism in planning, see Ole Jensen and Tim Richardson's (2004) book, *Making European Space: Mobility, Power and Territorial Identity*. For some more general thoughts on relationalism and governance see Andrew Barry's (2001) book, *Political Machines: Governing a Technological Society*. On environmental planning, Susan Owens and Richard Cowell's (2002) book, *Land and Limits*, is pretty comprehensive.

Notes

1. The material used in this chapter is derived, in the main, from the arena of planning theory. The decision to use theoretical reflections on planning process and practice rather than primary analyses of specific planning systems has been made so that the background assumptions that guide planning can be ascertained and analysed. It is in theoretical accounts that we can arguably see most clearly the spatial imaginaries at work in planning. Moreover, the virtue of using theoretical resources is that we do not need to confine the analysis to any specific national planning system: the following arguments can be applied to most national planning contexts; they concern the grounding principles of planning rather than any specific aspect of planning practice.
2. We see here evidence of the 'representing' and 'intervening' described by Hacking (1983) in the case of the laboratory sciences.
3. In many respects, Abercrombie's attachment to survey and plan is derived from the earlier work of Patrick Geddes. However, as Dehaene (2004) points out, the main distinction between the two is that Geddes sees survey and plan proceeding simultaneously while in Abercrombie's planning documents, survey and plan are normally organized in discrete stages.
4. As Daniel Smith (2003: 309) explains, an immanent process must function as 'a principle of critique as well as of creation [...] what "must" always remain normative is the ability to critique and transform existing norms, that is, to create something *new*'.

5. Following Flyvbjerg's (1998) analysis, outlined above, we need only think of the potential problems that would arise were the planners of Aalborg to introduce such a notion as environmental capacity into their city plans. It seems likely it would inevitably fall victim to the Chamber of Commerce's political manoeuvring.

6. I am here aligning 'ecological' with 'immanent', following comments in Thrift (1996: 28). The argument presented below suggests that were planning to fully engage with political ecology it would necessarily become more deeply immersed in heterogeneous relations of various kinds. It would become truly 'immanent' to the places being planned.

7

Dis/Ordering space III: the case of food

It is never simply a matter of speed [...] but of speed and slowness. There can be no acceleration without a parallel deceleration, no convergence without divergence, and no compression without decompression. (Doel, 1999)

Introduction

In the previous chapter, we examined the emergence of particular governmentalities in the arena of land-use planning. These governmentalities had developed on the basis of distinct spatial imaginaries, conceptions of the spatial realm that define legitimate planning actions and interventions. Two main spatial imaginaries were identified: first, topographical conceptions of well-ordered spaces in which entities are arranged by powerful technologies of planning (the map, the survey, the computer package and so forth); second, topological conceptions in which social, economic and technological processes are given free play so that discrete spaces emerge from the complex interplay of varied entity types. We saw that while planning often seeks to balance topographical and topological modes of ordering, it usually allows one spatial imaginary to dominate the other. We also observed that planning remains 'semi-detached' from space – that is, it fails to fully engage with the heterogeneous nature of relational space (as outlined in Chapters 2 and 3). In particular, it struggles to fully engage with the environmental and natural relations that remain so central to the spatial domain. We concluded that planning perhaps needs to develop a new spatial imaginary derived from political ecology, so that any planning interventions in spatial formations can be based upon the full range of entities normally found within such formations. The objective of planning should thus be to achieve a full integration of natural and social entities in the form of collectives.

In this chapter, we will focus more fully upon political ecologies of space by examining a spatial arena in which heterogeneous complexity is fully foregrounded: the food sector. Food is necessarily a mixture of the organic and the inorganic, the material and the symbolic, the social and the natural. Moreover, the food sector is also an arena in which vigorous efforts are being made to protect the natural components of foods against industrial 'substitution' (Goodman et al., 1987). Thus, numerous non-governmental groupings now strive to ban genetically modified foods, to promote organic foods and to support animal-friendly foods. In other words, ecological issues in the food sector are already politicized.

The politics of food takes on a necessarily spatialized character. Food is generally grown across extensive spatial areas. Likewise, food consumption takes places in differentiated cultural spaces. Thus, efforts to either industrialize or deindustrialize food must be played out in spatial terms. In general terms, then, we can trace the emergence of two contrasting food spaces: on the one hand, industrialized and standardized spaces that are subject to processes of continuous technological innovation in line with principles of economic efficiency; on the other hand, diverse local food spaces that are sustained by adherence to long-standing processes of production and consumption, processes that are deeply rooted in local cultures and natures. By differentiating these distinct food spaces – one broadly topographical, the other broadly topological – it is possible to see that patterns of development in the food sector are diverse and multiple rather than singular and uniform. However, we need to question whether the industrial and artisanal spaces that help to define 'foodspace' can easily co-exist. Although the contemporary food market may be able to accommodate (at least temporarily) the various commodities emerging from differing food networks, it is likely that contradictions between the production of large volumes and the production of distinctive and high-value foods will become ever more pronounced. For instance, industrialized foods challenge the conventional notions of quality that have long been established around traditional and 'natural' methods of production, while the reassertion of alternative foods implies a turning away from industrial technologies and a rediscovery of more typical or authentic production processes. In short, differing parts of the food sector appear to be heading off on opposing trajectories of development, some towards a more refined or intensive application of science and technology (for example, genetically modified foods), others towards a re-engagement with natural or traditional production methods (for example, organic and traditional foods). These divergent trajectories allow us to directly compare what we might call 'topographical spatial strategies' against what we might define as 'topological spatial strategies'. We can discern both in the food sector to the extent that a profound conflict between them is now becoming evident.

In this chapter we will examine the contested nature of food through the analysis of two contrasting food networks, chosen to represent the two food spaces identified above. The first is an archetypal illustration of industrialization and standardization: McDonald's. We show that this food chain is based upon a highly prescriptive set of relations that works to disseminate a uniform product, using a tightly controlled food delivery system. As we shall see, the prescriptive nature of the McDonald's system is derived in large part from the alignments of materials that connect food production and processing to the final consumption of the product. These materials facilitate the flow of a 'McDonaldized' cuisine into many diverse locales, allowing a truly globalized food network to emerge. We then turn to examine a second case study – Slow Food – which displays a markedly divergent set of connections between network and space. The Slow Food movement aims to reassert the value of the diverse local foods that can still be found in many local areas. Slow Food emphasizes that cuisines should reflect differentiated norms, practices and ecologies rather than some standardized model of food delivery. For Slow Food cuisine variation goes hand in hand with spatial variation. The network therefore works to strengthen the cultures and environments associated with local production and consumption practices. Thus, rather than binding local areas into a prescriptive set of uniform spatial relations, Slow Food promotes autonomy, fluidity and complexity within its own network spaces.

Having outlined these two contrasting food networks (one 'fast', the other 'slow'), and having shown that contemporary landscapes of food gain shape through differentiated network activity, we then move on to consider how distinct network spaces come to be constructed in line with specific conceptualizations of consumer practice. We consider two main approaches, loosely described as 'distraction' and 'engagement'. The process of 'distraction' ensures that the attention of consumers is drawn away from food towards other aspects of the food consumption 'experience', including the cost of food, the speed with which it can be prepared and consumed, and the eating environment. 'Engagement' can be seen as a reaction against such 'distracted' consumerism. It entails the consumer becoming strongly linked not just to the food itself but to the spatial contexts of food production. This linking can be achieved by the cultivation of 'taste' – that is, the physical ability to savour the various properties of food – as well as by the acquisition of knowledge about the cultural and ecological associations surrounding food production. The significance of this process of engagement encourages us to speculate that many consumers are now entering into a new commitment to food. This commitment apparently involves a heightened awareness of the ecological relations in which food is inevitably embedded. It is argued that a new concern for 'embeddedness' may indicate that a form of 'relational reflexivity' is emerging amongst discerning consumers. This form of reflexive consumption inevitably brings the relational nature of food to the forefront of consumer concern. Moreover, it shows how the assertion of spatial relations might become a key part of political strategies oriented to countering industrialization and standardization in the food chain. In short, it puts into action the politics of ecology identified by Latour (2004).

Spaces of fast food

We begin with perhaps the premier example of a globalized food network – McDonald's. While this company has long been emblematic of an industrialized and standardized food system, it nevertheless emerges at a particular time and place – the suburban US of the 1940s and 1950s – a place of rising wages, a boom in the birth rate, mass ownership of the motor car and increased leisure time. As Fine (1996) shows, this was a period when eating out became a standard pastime, with hamburgers and other fast foods increasing in popularity. Such foods, Rifkin (1992: 260) argues, met the new suburban requirement for 'convenience, efficiency and predictability in […] food preparation and consumption'. The hamburger, in particular, came to reflect prevailing cultural aspirations, its 'capacity for speedy preparation with uniformly satisfactory results […] meshing well with […] demands of consumer and entrepreneur alike' (Jakle and Sculle, 1999: 144). As Schlosser (2001: 60) summarizes it, 'a hamburger and french fries became the quintessential American meal in the 1950s'.

It was in this context that the McDonald brothers, Maurice and Richard, opened their first restaurant in Pasadena, California. The first McDonald's 'drive-in' sold mainly hot dogs to the new car-bound customers. After the success of this venture, the brothers moved to San Bernadino where they opened a bigger 'drive-in', which was even more successful than the first. However, the business was beset with problems, notably high labour turnover, so that in a tight labour market the brothers experienced recurring difficulties in recruiting new workers. In the late 1940s, they therefore closed the 'drive-in' and opened a new type of restaurant, one that was based on a less labour-intensive system of food delivery – what they came to call the 'Speedee Service System'. Under this 'System' the brothers,

> got rid of everything that had to be eaten with a knife, spoon, or fork […] [They] got rid of their dishes and glassware, replacing them with paper cups, paper bags, and paper plates. They divided the food preparation into separate tasks performed by different workers. To fill a typical order, one person grilled the hamburger; another 'dressed' and wrapped it; another prepared the milk shake; another made the fries; another worked the counter. For the first time, the guiding principles of a factory assembly line were applied to a commercial kitchen. (Schlosser, 2001: 20)

By employing an 'assembly-line' process, the McDonald brothers were able to diminish their labour requirements but could still deliver large quantities of burgers at low cost: 'a 1.6-ounce hamburger, 3.9 inches in diameter, on a 3.5 inch bun with .25 ounces of onion sold for 15 cents – a standardised product of high quality but also low price' (Jakle and Sculle, 1999: 141).

The popularity of this first McDonald's restaurant meant it attracted a great deal of attention, none more so than from a travelling milkshake mixer salesperson named Ray Kroc. On visiting the restaurant Kroc was immediately impressed by the efficiency of the operation. Ritzer (1993: 32) quotes him as saying: 'I was fascinated by the simplicity and effectiveness of the system […] each step in producing the limited menu was stripped down to its essence and accomplished with a minimum of effort. They sold hamburgers and cheeseburgers only. The burgers were all fried the same way'. After entering into negotiations with the McDonald brothers, Kroc took control of the McDonald's trademark in 1955 (the brothers had no desire to extend the system beyond the one restaurant). He then set about expanding the number of McDonald's restaurants across the US.[1] He opened his first restaurant in De Plaines, Illinois, close to the commuter station where he took the train to work. It was his hope that the new McDonald's would draw in a 'youthful, growing, and home-bound trade' – that is, the new suburbanites (Jakle and Sculle, 1999: 146).[2] In the following years, the number of restaurants increased rapidly, reaching 200 by 1960, making it the leading fast-food chain in the US.

Once the company was listed on the US Stock Exchange in 1965, profits had to be maintained through a relentless increase in the numbers of new restaurants. McDonald's thus moved away from its suburban roots and opened outlets in all the major US cities. This move was supplemented by expansion overseas as the company sought to maintain earnings growth. By the mid-1990s, it had 25,000 restaurants and operated in almost 120 countries (global earnings at this time stood at around $11 billion).[3] A key contributory factor to the growth of the chain was the introduction of franchising arrangements. Franchising is a system by which one firm sells the rights to distribute its products to a number of smaller firms. Kroc introduced this system into the McDonald's network in the late 1950s so that a large minority of the new restaurants became independently owned. In theory, this meant that the net-work comprised large numbers of autonomous restaurant spaces. However, at the same time as the franchising system was introduced Kroc acted to maintain central control. For instance, in 1958 he produced an operations manual to guide practices in all McDonald's restaurants. This manual was highly prescrip-tive, as the following comment makes clear:

> It told operators exactly how to draw milk shakes, grill hamburgers, and fry potatoes. It specified precise cooking times for all products and temperature settings for all equipment. It fixed standard portions on every food item, down to the quarter ounce of onions placed on each hamburger patty and the thirty-two slices per pound of cheese. It specified that French fries be cut at nine thirty-seconds of an inch thick. And it defined quality controls that were unique to food service, including the disposal of meat and potato products that were held more than ten minutes in a serving bin. (Love, 1986: 141–2).

The concern for uniformity led also to the creation of the 'Hamburger University' in 1961. The 'graduates' from this University (which by 2004 numbered around 65,000 persons (Ritzer, 2004: 41)) were expected to manage 'their' McDonald's restaurants in line with centrally formulated principles and practices. Thus, Kroc attempted to retain Panoptical oversight despite the increasing numbers of apparently autonomous spaces in the form of franchised restaurants.

According to Ritzer (2004, 40), it is the standardized character of McDonald's that has guaranteed its success: 'This uniformity allowed McDonald's to differ-entiate itself from its competitors, whose food was typically inconsistent'. In his influential book, *The McDonaldization of Society* (1993), Ritzer goes on to examine in some detail how McDonald's engineered standardized and uniform outcomes. For instance, he sees prescriptive mechanism at work in almost every aspect of the McDonald's food service system. He argues that these mechanisms are constructed out of nonhuman technologies that serve both to deliver food at a rapid pace and to regulate the actions of McDonald's workers. Schlosser also draws attention to such technologies in the following description of a McDonald's restaurant in Colorado Springs:

> Robotic drink machines selected the proper cups, filled them with ice, and then filled them with soda. Dispensers powered by compressed carbon dioxide shot out uniform spurts of ketchup and mustard. An elaborate unit emptied frozen french fries from a white plastic bin into wire mesh baskets for frying, lowered the baskets into hot oil, lifted them a few minutes later and gave them a brief shake, put them back into the oil until the fries were perfectly cooked, and then dumped the fries underneath heat lamps, crisp and ready to be served. Television monitors in the kitchen instantly displayed the customer's order. And advanced computer software essentially ran the kitchen, assigning tasks to various workers for maximum efficiency, predicting future orders on the basis of ongoing customer flow. (2001: 66)

The use of nonhuman technologies ensures that food preparation is simple and standardized: as the food arrives at the restaurant 'pre-formed, pre-cut, pre-sliced and "prepared" [so there is] usually no need [for the workers] to form the burgers, cut the potatoes, slice the rolls, or prepare the apple pie. All they need to do is, where necessary, cook, or often merely heat the food and pass it on to the customer' (Ritzer, 1993: 103).[4] Even the cash register has a simplified set of keys with labels such as 'Big Mac' or 'large fries' written on them. It is, therefore, 'not necessary for the cashier to know the actual price of any item, and the machines are programmed to "suggestive sell", so that dessert items, for example, will light up automatically to remind workers to suggest dessert to the customer who has not ordered it' (Fantasia, 1995: 208).

In the McDonald's system, then, highly prescriptive mechanisms work to regulate the system of food delivery. These mechanisms can also be found in the restaurants themselves, which are designed to exact specifications, wherever they might be. In fact, McDonald's goes out of its way to make the consumption experience as repetitive as possible, with symbols, signs, colours, layouts all repeating the basic formula: 'each McDonald's presents a series of predictable elements – counter, menu posted above it, "kitchen" visible in the background, tables and uncomfortable seats, prominent trash bins, drive through windows and so on' (Ritzer, 1993: 81). In other words, the various elements are drawn together to comprise a topographical space in which behaviour is rendered routine and predictable. The customer is expected to walk into the restaurant, queue, order, pay, wait a short time, take the tray, find a table, eat, put the rubbish in the trash can and (usually) leave. There may be some scope for lounging but, in the main, the restaurant is designed to speed the customer through a series of strictly co-ordinated interactions. As Ritzer (2004: 15) observes: 'lines, limited menus, few options and uncomfortable seats all lead diners to do what management wants them to do – eat quickly and leave'.

BOX 7.1

Spaces of fast food are:

- Configured by conventions of speed and efficiency, the need to ensure a constant flow of food through the preparation process and into the customer's mouth.
- Highly regulated according to some simplified but widely disseminated standards, notably in the structure of the food preparation process and in the layout of the restaurant spaces.
- Places where narrow ranges of food tastes are catered for as the standardized products are tailored to a mass market.
- Underpinned by heterogeneous networks in which nonhuman technologies are used to stabilize the conventions and regulations.
- Strongly prescriptive so that the behaviours of human actors are circumscribed by the alignments of materials. Prescription applies both to the employees in the restaurants who must 'fit in' to the standardized food delivery process, and to customers, who must abide by the main principles of fast food consumption ('eat and go').

Despite (or perhaps because of) its highly prescriptive character, McDonald's has been extremely successful in defining spaces of food consumption in many different places around the world. At the same time, the expansion of McDonald's has inspired a number of critical reactions, as in the case of the Hampstead residents in north London who sought to prevent McDonald's opening a restaurant in their up-market high street. Beyond this 'elitistic' criticism of McDonald's (which is largely based on the populist image of the restaurant chain) more substantive challenges have been mounted to the spread of the network. For instance, it is accused of producing food that is rich in sugar and salt, a dietary mix that is seen as contributing to obesity (Vidal, 1997; Critser, 2004). This has led to challenges in the courts in the US, on the grounds that consumers have received no warnings from the company on the likely health consequences of eating too much fast food. As *The Times* newspaper reported on 29 June 2004:

> A new litigation craze is sweeping America. Producers and sellers of food and drink are facing lawsuits claiming that they are responsible for obesity. The new growth industry has echoes of the crusade against the tobacco companies that has so far cost the tobacco

industry more than $250 billion [...] In the case of Pelman versus McDonald's Corporation (2002), a class action brought in New York state, consumers alleged that their obesity and diabetes were caused by eating McDonald's meals. The judge dismissed the case saying that the sale of food high in salt, fat and sugar could not be said to be inherently dangerous, given that such qualities were well known to the public. There was, however, a veiled warning to the food industry with the judge indicating that for such an action to be likely to succeed it would be necessary to show that the food was harmful in a way not known to the consumer. (Mullins and James, 2004: 6)

Although McDonald's has yet to lose a court case of this kind, it has already begun to respond to consumer concerns about obesity. For instance, it has recently announced the phasing out of its supersize range of meals and has begun to provide salads and other such 'health' foods in its restaurants.

While the obesity epidemic has the potential to change the composition of fast food, critics claim the main problem with the chain is the way it acts as an agent of standardization throughout the *entire* food chain. A number of commentators claim this has highly damaging side-effects:

Behind the front counters of the fast food chains, the familiar menus and logos of McDonald's [...] lie other assembly-line operations, ownership of which is concentrated in ever fewer hands, allowing even greater economies of scale. Thirteen large slaughter-houses, or meat packing houses in US terminology, now supply most of America's beef. Three companies, Simplot, McCain and Lamb Weston (which is owned by the even larger conglomerate ConAgra), control 80 per cent of the US market for frozen french fries. In the wake of the launch of the Chicken McNugget – made from constituted chicken, flavoured with beef additives and containing twice as much fat for its weight as hamburgers – eight chicken processors ended up with about two-thirds of the US market. (Meek, 2001: 3; see also Tansey and D'Silva, 1999; Schlosser, 2001)

The same process also affects the ecological realm, as Ritzer points out:

McDonaldization has produced a wide array of adverse effects on the environment. One is a side effect of the need to grow uniform potatoes from which to create predictable French fries. The huge farms of the Pacific Northwest that now produce such potatoes rely on the extensive use of chemicals. In addition, the need to produce a perfect fry means that much of the potato is wasted, with the remnants either fed to cattle or used as fertilizer. The underground water supply in the area is now showing high levels of nitrates, which may be traceable to the fertilizer and animal wastes. Many other ecological problems are associated with the McDonaldization of the fast food industry: the forests felled to produce paper wrappings, the damage caused by packaging materials, the enormous amount of food needed to produce cattle feed and so on. (2004: 17)

For its critics, McDonald's represents an extreme case of what we might call the 'will to simplify'. It has honed a number of procedures and techniques in the pursuit of a very narrow and particular set of aims (the most salient of which is 'to dominate the global foodservice industry' (McDonald's, 1996)). These procedures and techniques rely upon both human and nonhuman resources for their enactment. In short, McDonald's skilfully aligns heterogeneous elements to deliver a standardized cuisine (albeit with some limited variation to cater for local sensibilities; an arrangement that ensures predictable outcomes in diverse local circumstances. As Barry Smart (1994: 34) puts it, in the McDonald's network, 'irregularities of space and time are smoothed out under the market pressure of a remunerative uniformity'. Topology is subsumed within topography.

Spaces of slow food

As mentioned above, opposition to McDonald's is extensive and takes a variety forms, including direct attacks on restaurants during anti-globalization demonstrations. However, our second case study represents a more long-standing and constructive form of opposition to McDonaldization in the food sector. It concerns Slow Food, a consumer movement that was established in Italy during the mid-1980s in direct response to the opening of a McDonald's restaurant in the famous Piazza d'Espania in Rome (see Resca and Gianola, 1998, for a full account of this controversy). The opening of a McDonald's in this prestigious location raised the possibility that traditional Italian eating habits might be under renewed threat from 'Americanized' fast food. As part of the ensuing protest, the food writer Carlo Petrini initiated a meeting of chefs, authors, journalists and other intellectuals to discuss the most effective means of protecting traditional Italian cuisines from foreign invasion. This first meeting gave birth to a new consumer movement – Slow Food – which was to be devoted to the promotion of an 'anti-fast food' culture. As Renato Sardo, the director of Slow Food International, puts it:

> There was a lot of public debate at the time about standardisation, the McDonaldisation, if you will, of the world. Up until then, any opposition was split in two. On the one hand there were the gastronomes, whose focus was fixed entirely on the pleasure of food. The other tradition was a Marxist one, which was about the methods of food production and their social and historical implications. Carlo Petrini, Slow Food's president, wanted to merge the two debates to provide a way forward. (quoted in *The Observer*, Food Monthly, 11 November 2001)

In meshing these two sets of concerns, Slow Food sought to bring discerning 'gastronomes' to the rescue of traditional foods (the middle class would come to the aid of the peasantry). This would involve the targetting of discriminating consumers in order to heighten awareness of 'forgotten' cuisines. By this means, it was hoped that a new market for traditional local foods could be created.

The main device for reaching potential consumers of 'slow' products was to be a new publishing house (Slow Food Editore), which would disseminate informed, interesting and accessible material on previously unknown or neglected foods. It was also intended that a network of local groups would be established in order to identify foods that are central to local cuisines. According to Carlo Petrini (2003: 12–13) the groups would base their activities on four major themes:

FIGURE 7.1   Slow Food's symbol of 'slowness'. Illustrator Andrea Pedrazzini.
(Source: Slow Food, 1999)

1. The study of material culture: 'This is the movement's principal theoretical and behavioural guideline: namely, that it is pointless to sing the praises of fine wine or the smell of good bread if you don't know how they are produced'.
2. Preservation of the agricultural and alimentary heritage from environmental degradation: 'the organoleptic profile of the food we eat (in other words, how it strikes our sensory organs) is being constantly impoverished. If that doesn't deserve high quality production what does?'.
3. Protection of consumers and producers: 'letting people know, without rhetoric or bombast, where to find the right combination of quality and price, neither praising things that are good but expensive nor those that are cheap but substandard'.
4. Promotion of the pleasures of gastronomy: to be conducted 'in a genial and tolerant manner that encourages an approach to food based on the hedonistic advantages of deeper knowledge, the education of the senses, and harmony around the table'.

The first edition of the movement's magazine, *Slow*, upheld these aspirations but also showed that Slow Food was explicitly seeking to oppose the spread of McDonald's and the other fast-food chains. Slow claimed that the organization stands in opposition to the 'folly of fast life' (Petrini, 1986); it proclaimed the need to nurture 'gentleness, pleasure, knowledge, care, tolerance, hedonism, balsamic calm, lasting enjoyment [...] culinary traditions' (Petrini, 1986). The symbol of the snail was adopted as the movement's logo. For Carlo Petrini:

it seemed [...] that a creature so unaffected by the temptations of the modern world had something new to reveal, like a sort of amulet against exasperation, against the malpractice of those who are too impatient to feel and taste, too greedy to remember what they had just devoured. (1986: 1)

As the adoption of this symbol suggests, the emphasis in Slow Food is on *time* and on the need to decelerate the food consumption experience so that new (or, perhaps more accurately, *old*) forms of taste can be re- acquired. As Alberto Capatti (1999), a leading member of the movement, puts it: Slow Food 'is profoundly linked to the values of the land and the past. The preservation of typical products, the protection of species from genetic manipulation, the cultivation of memory and taste education – these are all aspects of this passion of ours for time'. However, Slow Food also has spatial significance: the movement is concerned by the rupture that has been effected between spaces of production and spaces of consumption, and it seeks to close the gap between the two by bringing consumers to spatially embedded foods. It also wishes to reassert the natural bases of food production (seasonality, ecological content, etc.) and the role of cultural context (tacit knowledge, culinary skills, etc.). In short, it wishes to re-embed food in topological complexity.

Slow Food's main concern is for 'typical' or 'traditional' foods. According to Torquati and Frascarelli (2000: 343), 'typicality' in the Italian context is determined by: 'historical memory (the product is associated with the history and with the traditions of the place of production), geographic localisation (influence of the pedoclimatic environment), [and the] quality of raw materials and techniques of preparation'. However, while it attempts to bolster these components of local cuisines, Slow Food recognizes that local and regional food products are disappearing because they are *too* embedded in local food cultures and ecologies; they are not easily extracted and sold into modern food markets (either for cultural or ecological reasons they often cannot travel the long distances covered by McDonald's burgers). So the movement works to attract consumers to these traditional products by emphasizing the foods' aesthetic qualities. For instance, Slow Food's quarterly magazine, *Slow*, promotes a highly aestheticized form of consumption in its lavishly illustrated articles. Slow Food Editore also produces a number of glossy food guides that give consumers information on 'slow food' outlets. The most well-known of these guides, *Osterie d'Italia*, identifies typical restaurants in all the Italian regions, thereby giving new consumers (for example, tourists) the opportunity to engage with previously hidden, but long established, local foods.[5]

While its roots are very firmly within Italian food cultures, Slow Food seeks to promote typicality much further afield. In line with this goal, in 1989 it formally launched itself as an international movement and has subsequently spread to around forty countries (at the time of writing, it has around 70,000 members world-wide and now maintains offices in New York, Paris and Hong Kong). The many local Slow Food groups are organized at the regional level into 'convivia'. Although the bulk (300 plus) are located in Italy, convivia are now

L'*Arca* del *Gusto*
e i *Presidi*

Slow Food

FIGURE 7.2   Slow Food's Ark of Taste (Source: Slow Food, 2000)

operating in such differing contexts as Australia, Brazil, India and the US. Essentially, a convivium is a consumer club made up of people who wish to 'cultivate common cultural and culinary interests' (Slow Food, 1999). The definition of the local convivium area is given by cultural and culinary distinctiveness so that each convivium is charged with promoting a particular local cuisine. The convivia usually undertake the following activities:

1. identifying restaurants that enshrine the principles of 'slowness' (mostly those offering a good selection of regional dishes and wines);
2. organizing tastings of typical foods and talks by speciality producers and others on gastronomic issues;
3. promoting an appreciation of local foods in schools and other public institutions; and
4. highlighting the culinary creativity and tacit knowledge that reside in local cuisines.

In short, Slow Food seeks to build up cultural diversity by establishing close associations between local cuisines and local systems of production. It therefore conjures up collectives in which the interactions between both people and nature are intense and close. In seeking to strengthen these close associations, the movement has come to recognize that many (ecologically-embedded) producers and processors are precariously connected to consumers. It has therefore decided to initiate more direct action in the production sector through a scheme called the Ark of Taste (Slow Food, 2000). Along with the usual activities oriented to the dissemination of knowledge about endangered products, the Ark project aims to set up another local group structure ('Praesidia'), which will encourage people to locate producers in need of support, identify the most appropriate support measures (for example, new marketing channels), and raise funds in order to put these measures in place. In giving such support to producers, Slow Food explicitly intends to promote ecological diversity in local food spaces.

BOX 7.2

Spaces of slow food are:

- Arenas of diversity in which locally-embedded products are supported and promoted.
- Oriented to the development of 'taste' – that is, they encourage consumers to investigate new flavours and textures and to acquire some knowledge about the local provenance of the food ('taste education').

Continued

- Comprised of interactions between the local and the non-local as consumers are drawn into discrete cultural contexts of consumption.
- Areas in which local ecological conditions are sustained by the production of locally-specific food products.

The Slow Food network thus provides a useful comparator to McDonald's. Where McDonald's imposes a standardized format upon each locality, Slow Food encourages and supports multiplicity; where McDonald's is based upon the dissemination of a simple formula ('The Speedee Service System'), Slow Food is built on an appreciation of diverse food production processes (often of an artisanal nature). But more than this, there is a recognition in Slow Food that local diversity cannot simply be asserted as an alternative: first, local cuisines and their constituent products must be rendered transparent and made available to a wide number of potential consumers (this is done through the Slow Food publishing house); second, these cuisines must be protected by creating links between producers and consumers (this activity takes place through the 'convivia'); and third, producers must be enabled to remain in existence (such enabling efforts are made under the 'Ark of Taste?). All these activities require the construction of a network that extends from the local to the global, so that the varied cultures and ecologies found in Slow Food spaces can be both protected and supported.

Spaces of consumption

The previous section describes two contrasting but co-existing culinary networks: one is configured by the need to disseminate standardized food products in a mass-market; the other is shaped by the desire to enhance consumer access to the diverse products still to be found in many local areas around the world. Although these are both spatially extensive ('global') networks, they act to combine the local and global in very different ways: one attempts to make the local a mirror of the global (when you stand in one McDonald's, in a sense, you stand in them all); the other seeks to sweep up multiple localities into a loosely constructed and fluid global network, what Rajchman (2001: 55) calls (after Deleuze) a 'vagabond' set of relations.

In the preceding section, we assessed these networks from the perspective of the 'network builders' – that is, we examined how McDonald's and Slow Food work to extend their influence through space and time. However, in order to gain a better understanding of the interaction between each network and differing spaces of food, it is also worth examining how they work to draw consumers into their own relational configurations, for, ultimately, both are oriented to expanding the numbers of people consuming the products they promote.

Thus, in this section, we will assess the kinds of consumers that are 'constructed' by the two networks. As we shall see, these constructions are important not only because they serve to further distinguish the two networks but also because they indicate the character of contemporary food consumption and its association with relational space.

To begin once again with McDonald's, we can suggest that the company attempts to recruit consumers by making itself both widely available and readily accessible. As the journalist Martin Plimmer notes: 'There are McDonald's everywhere. There's one near you, and there's one being built now even nearer to you' (quoted in Ritzer, 2004: 2). Not only is McDonald's readily available, it is also easily accessible, as this comment by Gottdiener indicates:

> Fast food outlets are successful because they offer an easy solution to the method of purchasing food that depends little on spoken language, on the interpretation of the menu or personal relations with the waitress/waiter, as happens in other restaurants. These and other themed environments, with their over-endowed, instructive sign systems are fun places to be because they minimise the work we need to do for a successful interaction. (1997: 132)

Boym makes a similar point when he asks:

> What is it about McDonald's that attracts children and immigrants alike? As a rule, immigrants, like children, are very sensitive creatures. In their desire to blend in, they are conscious of making the wrong gesture, looking funny or different, standing out in any conspicuous way. The simple experience of entering a restaurant, asking for a table, and talking to a waiter can be intimidating. In this respect, McDonald's is the ultimate populist place. No one can be excluded, you can come and go as you please. It's okay to bring your children and to make a mess. Toys are given away along with nutritional information: there is something for everyone. (2000: 1)

Availability and accessibility are buttressed by the relentless efforts that McDonald's puts in to branding the company and its products. Through advertising and other forms of publicity, consumers already feel some familiarity with the brand even before they step into a McDonald's restaurant.[6] As Schlosser (2001: 5) notes: 'customers are drawn to familiar brands by an instinct to avoid the unknown. A brand offers a feeling of reassurance when its products are always and everywhere the same'. But more than this, the branding of McDonald's works to draw consumers into some form of symbolic relationship with the company. Elspeth Probyn (2000: 35) focuses on this aspect when she suggests that branding is aimed at constituting McDonald's consumers as a 'global family', where 'the Big Mac preceded the internet in bringing us all together [...] extending an ethics of care into the realm of global capitalism and creating its consumer as a globalised familial citizen'. This notion of the 'global family' works to consolidate a form of 'global relationalism', in which consumers of various kinds owe some kind of allegiance to McDonald's – they are part of an on-going and intimate relationship between the company and its customers.

Yet, what is striking about the relational belonging fostered by McDonald's is that it pays so little attention to the food sold by the company. In fact, the effort McDonald's makes to 'personalize' itself may actually require the *suppression* of knowledge about the processes of production that lie behind the restaurants' food products. This point is well made by Kroker et al., who argue that,

> [h]amburgers […] have been aestheticized to such a point of frenzy and hysteria that the McDonald's hamburger has actually vanished into its own sign. Just watch the TV commercials. Hamburgers as *party time* for the kids […] as *nostalgia time* for our senior citizens […] as *community time* for small town America and, as always, hamburgers under the media sign of *friendship time* for America's teenagers. (quoted in Smart, 1999: 13, emphasis in the original)

In short, the typical McDonald's consumer is confronted by layers of symbolic artifice. These layers surround the presentation of food in the restaurants and work to distract attention from the nature of the food itself.[7]

For Slow Food, however, quite another role is envisaged for the consumer. The key requirement here is not distraction but 'attentiveness', as Carlo Petrini explains:

> attentiveness to the selection of ingredients and the sequence of flavours, to show how the food is prepared and the sensory stimuli it gives as it is consumed, to the way it is prepared and the sensory stimuli it gives as it is consumed, to the way it is presented and the company with whom we share it. There are endless degrees of attentiveness, which in our view are just as important whenever and wherever we take nourishment, whether it is a meal at home or in a restaurant, a drink in an *osteria* or a sandwich at a bar, lunch in a school cafeteria or in an airplane. The real difference in quality among these experiences does not lie in how much time is devoted to them, but in the will and the capacity to experience them attentively. (2003: 33)

Importantly, the attention that Slow Food demands in the consumer extends beyond the meal into the cultures and ecologies of production. As Petrini (2003: 15) emphasizes, Slow Food requires 'an alert consumer, filled with curiosity, who [wants] to take part at first hand and learn'. This process of learning should ideally take the consumer into the heterogeneous relations that lie behind the product so that the cultural and environmental contexts of production are fully appreciated. As Petrini (2003: 69) puts it: 'we need to reconstruct the individual and collective heritage, the capacity to distinguish – in a word taste'. Here, taste is both a physical and a cultural attribute, it brings to the fore 'sensory experience', so that all the senses are working in the appreciation of food quality. As Petrini puts it: 'Pleasure is physiological' (quoted in *The Independent*, 11 October 2004).

BOX 7.3

Differing food spaces 'construct' differing consumers:

- Fast-food spaces 'construct' distracted consumers, who know little about the (standardized) products and are drawn to the strongly aestheticized arenas of consumption through the brightly coloured restaurants and widespread marketing campaigns.
- Slow food spaces 'construct' attentive and engaged consumers, who are encouraged to build relationships of various kinds to the diverse arenas of food production and preparation. These relations are 'gastronomic' in the sense that they are built upon knowledge about products and practices as well as 'sensory' connections to the food and its ingredients.

We can thus speculate that Slow Food is aiming to open up spaces of taste in which the senses are linked, through the food, to key aspects of the local ecology. It forges 'lines of flight' which extend from the mouth to the field. Culinary space thus comprises a space of sensory interactions between the consumer's physical ability to savour food and the ability of the local food system to deliver savoury food products. These 'lines' are both physical and cultural: they require an alignment of knowledge and practice. Moreover, Slow Food suggests that by opening the senses to physical tastes, and by understanding a little more about the origins of those tastes in cultural and ecological terms, consumers can 'resist McDonaldization'. Thus, Slow Food is opposed to fast food but its strategy is 'not so much a question of fighting a fundamentalist war' as it is of 'informing, stimulating curiosity, giving everyone the opportunity to choose' (Petrini, 2003: 69).

In stimulating curiosity, Slow Food aims to immerse consumers in new worlds of belonging. It aims to build new sets of relations in the food sector, relations that tie consumers more intimately to the cultures and environments of production. These are not the globalized relations of the 'McDonald's family'; rather, they are localized relations, which vary in line with the differing cultural and ecological conditions found at the local level. Thus, in opposition to the single (global) space fabricated by the fast food chains, Slow Food conjures up a host of multiple and fluid spaces, all of which hold their own distinctive cultural and ecological characteristics. Ultimately, Slow Food asks that consumers take the time to (slowly) immerse themselves in these spaces so as to appreciate the heterogeneous complexity to be found there. Through their senses and through the mobilization of cultural resources, consumers can enter into a new relationship with food space. As Parasecoli (2003: 30) summarizes it, in this food space 'pleasure, liminally situated between the symbolic and the biological, is considered liberating and disruptive, a primal force that can shake every structure to its base'.

Risk and relationality

In the previous section, we examined the way spaces of food are constructed as spaces of consumption. We suggested two main types of consumption – 'distracted' and 'engaged' – and demonstrated that these are promoted by the two networks identified earlier – McDonald's and Slow Food. Each network configures spaces of consumption in ways that promote the two forms of consumer behaviour. Thus, McDonald's constructs consumption spaces that are prescriptive in form but which come shrouded in symbols of various kinds. This leads to a 'personalization' of the food chain and even to the notion that McDonald's is a 'family' in which all consumers can be somehow related. Yet, the position of consumers in the McDonald's network is relatively fixed: once inside the restaurants they simply ingest the food quickly and then leave; the transaction between McDonald's and its users is simple and functional. Slow Food, on the other hand, endorses restaurants and other consumption spaces that enable consumers to expand their culinary knowledges and tastes. It thus works to consolidate food practices that are based upon diverse cuisines set within diverse cultural and ecological contexts. Here consumers can be shaped and modified by their immersion in 'slow' spaces. In this sense the network and its consumers co-evolve – tastes are developed and cuisines are strengthened as both come into some kind of alignment.

At this point, it is worth noting that the treatment of the two networks has so far been symmetrical – that is, we have characterized each as being a significant component of the contemporary food sector. However, the attentive reader might object that the symmetry between the two is misplaced for the 'distracted' consumers of the McDonald's network clearly outnumber by some considerable margin the 'engaged' consumers of Slow Food.[8] It might, therefore, be suggested that pitching these two networks against each other simply distracts us from acknowledging that the consumption of fast food is of substantially greater importance than the consumption of slow food.[9] While such observations are obviously well-founded, in this section, it will be argued that the significance of 'slow' consumption may be greater than the discussion of Slow Food has so far led us to believe. In particular, we will suggest that growing numbers of consumers are becoming more relational in their appreciation of food and that this relationality may ultimately lead to a significant reshaping of food space.

In part, the growing significance of 'slow' consumption can be attributed to food scares which alert even McDonaldized consumers to the product that lies 'behind' the sign (Beardsworth and Keil, 1997). Franklin, in reference to McDonaldized meat products, suggests that:

The new food scares did something which put into reverse one of the key characteristics of meat eating in modernity. They emphasised to the consumer, the connections between animals and meat, and underlined the process of animal-into-meat. These rationalised, intensive processes, so studiously hidden from the public gaze, were revealed to be the source of a new risk. In short, the new methods of meat production rendered all meat a potential health risk and it lost its innocence as a marker of modern progress. (1999: 164)

For Ulrich Beck (1992), this loss of innocence can be interpreted as symptomatic of a new modernity, one that he calls 'risk society'. Here, 'many things that were once considered universally certain and safe and vouched for by every conceivable authority turn [...] out to be deadly [for example, beef]' (Beck, in Slater and Ritzer, 2001: 293). Beck suggests that in this uncertain social context consumers are forced to become more 'reflexive' in their relationships with a whole range of goods, in part because they can no longer rely on expert institutions. In line with this view, Halkier (2001: 208) believes that individuals are currently being 'pulled between an increased insecurity about knowing what to do and an increased awareness of possessing agency, the capacity to do something'.

One means of resolving this tension between insecurity and capacity is through the conscious assessment of 'quality' (Harvey et al., 2004). As Slow Food emphasizes, if consumers are to make informed assessments of food, they require an awareness of the economic, social and ecological relations that underpin food production and manufacturing processes. Enhanced reflexivity around product quality may therefore prompt the emergence of a deeper understanding of the complex associations that inevitably surround food commodities. In the view of some commentators, there are good reasons for believing such an understanding may be emerging at the present time. David Goodman (1999), for instance, suggests that the food sector has now entered an 'Age of Ecology' wherein the complex 'metabolic reciprocities' that link production and consumption have come more fully into view (see also FitzSimmons and Goodman, 1998). This 'Age of Ecology' can be discerned, Goodman suggests, in the popularity of organic foods, which are held to retain key natural qualities, and in the consumption of typical and traditional foods, which are believed to carry cultural qualities associated with long-established cuisines. In their different ways, he argues, these food types challenge the instrumental rationalities of the industrialized food sector and imply the need for more relationally embedded forms of production and consumption.

Goodman's account seems to imply that consumers, in assembling food preferences, choices and tastes, are entering into a changed relationship with the objects of these preferences, choices and tastes. And in this changed relationship, they not only 'reflect' upon the qualities of food goods but express a desire to genuinely immerse themselves in natural and socio-cultural relations. Thus, organic foods promise some reconnection with a nature that is being increasingly lost to industrial foods, while traditional or typical foods promise a reconnection with social and cultural formations that were previously distant in space or time. By consuming such foods, consumers seem to aspire to a greater sense of connectedness in the hope that this connectedness will keep at bay the risks associated with industrialized foods (Nygard and Storstad, 1998).

BOX 7.4

Relational consumption arises:

- Because the advent of modern food scares (such as BSE) leads consumers into a greater awareness of the processes of production that 'lie behind' the products.
- This coincides with a 'turn to quality' amongst certain (discerning) consumers, so that locally embedded foods – organic foods, typical foods, fairly traded foods – become more popular. These foods promise a closer relationship between the consumer and the economic, social and ecological contexts of food production. They therefore promote relationalism.
- This move into relational consumption requires the consumer to exercise some critical judgement when buying products (what to buy and what not to buy) and also requires new connections to be made to particular 'lines of flight' out of the product – that is, the organic, typical, fairly traded pathways that link the product to its arena of production. This might be termed 'relational reflexivity'.

In assessing these two aspects of 'embedded consumption' – 'reflection', on the one hand, and 'immersion', on the other – we might follow Scott Lash (1998) in proposing that consumers need to balance 'experience' and 'judgement': that is, they need to apply an instrumental rationality (concerned, for instance, with risk or economic calculation) at the same time as they attempt to deal with indeterminacy and uncertainty in both the knowledge systems that underpin this rationality and in the goods themselves. Lash argues that the need to combine these two aspects of consumption practice will lead consumers to rely upon a new form of 'aesthetic judgement', one that involves both intellectual reflection (in order to establish a rule, something to guide the act of consumption) and imagination, understanding and feeling (in order to establish an aesthetic relationship with the commodity).

The concept of 'aesthetic judgement' proposed by Lash has something in common with Crang's (1996) notion of 'aesthetic reflexivity'. Crang suggests that such reflexivity entails tracing the emergence of food commodities as they move through spaces of production, processing and consumption. In Crang's (1996: 51) view, this approach involves 'roughing up the surfaces' of normally 'smooth', 'unblemished' commodities to reveal the webs of connection and association that necessarily compose food products (see also Bell and Valentine, 1997). An illustration of such aesthetic reflexivity is provided by Probyn (2000: 14) when she writes that a reflection on eating 'can be a mundane exposition of the visceral nature of our connectedness and distance from each other, and from our social environment'. It allows us to consider 'what and who we are, to ourselves and to others, and can reveal new ways of thinking about those relations [...] In eating, the diverse nature of where and how different parts of ourselves attach to different aspects of the social comes to the fore and becomes the stuff of reflection'. Probyn's discussion of McDonald's, vegetarianism, eating disorders and other aspects of the consumption process can be read as an attempt to reflect upon connectedness. It might therefore be seen as an attempt to utilize the notion of aesthetic judgement outlined by Lash and others. In Probyn's account, this aesthetic appears to have a dualistic quality. On the one hand, consumers must assess risks and other dimensions of the act of consumption in reflexive terms. Such reflexivity requires that a 'critical distance' is established between the subject and object of consumption so that an objective evaluation can be carried out. On the other hand, it requires a new aesthetic relationship of some kind so that a sensual connection, something that lies outside formal systems of calculation, can be established. By combining these two aspects, we can suggest that a form of 'relational reflexivity' comes into being (Murdoch and Miele, 2004) and that this provokes consumers into a new awareness of themselves as the subject and food as the object of the act of consumption.[10]

Conclusion

In the previous sections we have shown the landscape of food is constructed and consolidated by networks of differing kinds. We examined two such networks in some detail. First, McDonald's, as a global restaurant chain, distributes a standardized and ubiquitous product via an industrial system in which non-human technologies play a key role in both prescribing the actions of workers and in delivering food to the mouths of consumers. While it is willing, to a limited degree, to tailor its products to local circumstances (for example, selling pizzas in Italy, few meat products in Muslim countries, luxurious restaurant fittings in Monte Carlo), its strength is based on a standardized mode of food delivery. This extends throughout the McDonald's chain and beyond, to its many suppliers around the world. Second, Slow Food is a consumer network that works to promote diversity in food production and consumption processes in order to safeguard local cuisines. It therefore seeks to highlight the connections that link cuisines to local natures and cultures, and works to strengthen the markets for locally embedded products. Within 'slow' spaces consumers are expected to link their tastes to specific cultural and ecological materialities so that the entire culinary assemblage is strengthened.

These two networks give rise to contrasting spaces of food. McDonald's applies a uniform set of principles and seeks to turn all its network spaces into an expression of the same thing ... 'McDonald's'. This standardized space is engineered using Panoptical principles which are enshrined in management practices, work routines, physical arrangements, nonhuman technologies and so forth. Space in the McDonald's network is thus primarily topographical in character. Alternatively, Slow Food ties together a whole host of cuisines within sets of relations that give aesthetic expression to spatial diversity. It links local areas within a loosely consolidated assemblage. These local areas retain distinctive socio-material relations that are sensitively 'globalized' in the Slow Food network. Thus, Slow Food reminds us that, as Massey puts it,

> local places are not simply always the victims of the global [...] places are also the moments through which the global is constituted, invented, coordinated and produced [...] this fact of the inevitably local production of the global means that there is potentially some purchase through 'local' politics on wider global mechanisms. Not merely defending the local against the global, but seeking to alter the very mechanisms of the global itself. A local politics with a wider reach; a local politics on the global. (2004: 11)

Enhanced understanding of the differing relations between global and local spaces in McDonald's and Slow Food gives rise to a need to critically evaluate the networks one against the other. Clearly any evaluation could begin to think about their differential impacts on landscapes, cultures and ecologies. It might, for instance, point out that McDonald's seems to 'externalize' many of the most significant interactions between food and environment by drawing producers, processors and consumers together within narrowly defined instrumental relationships that are dominated by industrial and market conventions (Murdoch and Miele, 2004). As a consequence, Schlosser (2001: 261) claims 'the low price of the fast food hamburger does not reflect its real cost', while other aspects of production/consumption are displaced by McDonald's (notably the health effects of the fast diet (see Vidal, 1997; Critser, 2004), so that 'the profits of the fast food chains have been made possible by the losses imposed on the rest of society' (Schlosser, 2001: 261). Slow Food, on the other hand, appears to encourage some 'internalization' of these costs within economic processes. In this network, the full range of spatial consequences of production and consumption are assessed. But this 'internalization' of cost means that many of the 'slowest' foods are relatively expensive. Thus, the Slow Food approach requires an 'aestheticization' of typical foods in order to attract those consumers who are willing to look beyond price to a much broader set of criteria. The apparent success of Slow Food in establishing an extensive global network based on an aestheticized food culture seems to indicate that, at present, a growing number of consumers are disposed to assessing food in this fashion. The increased significance of 'slow food' may therefore point to an enhanced relational awareness in food consumption practices more generally. In short, modern consumers may be starting to acquire a topological sensibility so that they routinely look 'beyond' the product to the complex relations that comprise economies, cultures and ecologies of production and consumption.

SUMMARY

In this chapter we have looked closely at two competing food networks. It has been suggested that these two networks give rise to differing food spaces. In the first, McDonald's, a rather prescriptive and simplified space comes into being, oriented to the speedy and efficient delivery of food. In the second, Slow Food, a fluid and complex space emerges in which food is seen as a way of promoting relational attachments to territory and culture. These two networks mobilize two differing types of consumer: the first is 'distracted', the second is 'engaged'. While the McDonald's fast-food space perhaps dominates at the present time, it was argued that there are grounds for thinking that 'slowness' may be gaining ground.

FURTHER READING

For a general introduction to geographies of food, see David Bell and Gill Valentine's (1996) book, *Consuming Geographies: You Are Where You Eat*. Readers wishing to investigate the McDonald's network should turn to George Ritzer's (2004) book, *The McDonaldization of Society* (Revised New Century Edition). This provides a synthesis of many writings on fast food. For those wishing to acquaint themselves with 'slow food', see Carlo Petrini's (2003) reflection on the emergence of the Slow Food movement, *Slow Food: The Case for Taste*. For a Deleuzian approach to food consumption issues, see Elspeth Probyn's (2000) text, *Carnal Appetites*.

Notes

1. Kroc finally took over total control from the McDonald brothers in 1961 for $2.7 million. After this time, he was free to build the business as he saw fit (Ritzer, 2004).
2. According to Jakle and Sculle (1999), it was not simply the rationalized food delivery system that appealed to Kroc: he also saw the restaurant as emblematic of a 'suburban lifestyle', one that he himself aspired to.
3. By 2002, McDonald's total sales had reached over $41 billion and the chain comprised over 30,000 restaurants, over half of which could be located outside the US. The chain now serves around 46 million customers each day (see Ritzer, 2004).
4. Cooking, however, is also conducted by technological components as Fantasia (1995: 208) points out: 'the food in a McDonald's outlet is prepared by the use of timing mechanisms, beeping signals, pre-measured quantities, and computers submerged in the cooking oil that fry foods to uniform specifications'.

5. The publication of such guides enables Slow Food to run its own version of the (McDonald's) franchising system: independently owned and run restaurants that promote local and traditional cuisines are all brought under the Slow Food umbrella.

6. Branding also assists familiarity because of the way it is embedded in the restaurant space. For instance, Gottdiener (1997: 81) shows that McDonald's 'capitalises on the many thematic elements it has produced in advertising over the years, such as the cartoon characters associated with Ronald McDonald and his friends'. These elements are used to 'theme' the restaurants so that the restaurant functions as a 'total thematic environment'.

7. This notion of 'distraction' is given further credence by Fiddes's analysis of blood in meat:

> Meat [...] is intrinsically linked with red blood – but the colour carries with it a series of associations largely concerning power, violence and danger. These are ideas which the fast food purveyors would be keen to curtail, since part of the burger's attraction is its sanitised supply. Instead, pastel shades represent a gentler image than that of fiery, savage red. Sometimes even green is employed – the colour of chlorophyll – which stands increasingly for nature, for health, for freshness. (Fiddes, 1991: 116)

8. Also, as Ritzer points out:

> in an increasing number of American communities, there are fewer alternatives to McDonaldized settings. Even if young people didn't want to eat in such settings (and they usually do!), they are increasingly forced to by the absence of alternatives. The result is that by the time they reach adulthood, a growing number of Americans have experienced relatively few non-McDonaldized settings. The fast food restaurant and its products are the standards by which younger Americans judge the alternatives. Thus, to many of them, a home cooked gourmet hamburger is not likely to taste as good as the McDonald's hamburger they have been exposed to, and have eaten, all their lives. The McDonald's hamburger has become the standard of quality and against that all alternatives are likely to be judged negatively. (1999: 20)

9. This is especially so if the rapidly expanding market for ready meals that simply need to be placed in a microwave for a few minutes is taken into account.

10. There is a neat parallel here with Deleuzian discussions of art. Peter de Bolla, in a review of Deleuze's (2004) book *Francis Bacon: The Logic of Sensation*, makes the following point:

> The 'logic of sensation' is peculiar to art; it is one of the ways in which art makes its difference from philosophy felt. The philosopher sets out to create or invent concepts; the painter aims to 'paint the sensation' [...] This doesn't mean there can be no philosophical interest in painting. Far from it: since art's logic of sensation is an example of transcendental empiricism, it enables me to distance myself from my own senses or my sensation of sensing the world, thereby forcing me to invent a different way of conceiving of myself as the subject in and object of experience. (de Bolla, 2004: 20)

8

Post-structuralist ecologies

[I]t is not a question of anti-humanism, but a question of whether subjectivity is produced solely by internal faculties of the soul, interpersonal relations, and intra-familial complexes, or whether non-human machines such as social, cultural, environmental assemblages enter into the very production of subjectivity itself. (Goodchild, 1996)

Introduction

In the preceding pages, we have charted a course across the rocky terrain of post-structuralist geography. We have taken in landscapes of fluidity and instability, as well as landscapes of permanence and solidity. We have encountered spaces of discipline and confinement, as well as spaces of movement and transformation. We have analysed heterogeneous associations, as well as the spatial imaginaries that animate such associations. We have reviewed the metaphorical terms used by post-structuralists to describe space and place and, in so doing, we have engaged with processes of network building, processes of emergence, processes of stabilization, processes of division, processes of *de/re*territorialization and so on and so forth.

Some of these post-structuralist geographies were explored in the case-study chapters, which concentrated in the main upon the various ways in which spatial relations interact with spatial locations. In the first case-study chapter, we investigated this interaction in the domain of 'nature' and saw that efforts to secure nature within clearly demarcated spatial zones inevitably lead to the surfacing of heterogeneous relations. These relations refuse to respect zoning operations; rather, they perform transgressive movements which undermine on-going efforts to shore up spatial defences. Although this finding implies a requirement to abandon strategies of spatial purification (that is, spaces *for* nature and spaces *for* society), it also indicates that a new interaction between spatial relation and spatial location could and should be generated, especially if natural entities are to be effectively stabilized in dynamic ecological contexts.

In the second case-study chapter, it was suggested that any full engagement with nature requires a shift in spatial imaginaries. Using the example of planning (notably planning theory), it was argued that physical, social and political imaginaries must be recast and recombined in order to allow the development of an 'ecological imaginary'. This 'ecological imaginary' would bring into being new collectives of heterogeneous entities. Within these collectives, 'partnerships' between humans and non-humans would be stabilized in ways that allow both 'types' to bolster the integrity of the other. This requires attention to the general ecological conditions enjoyed by the collective, the distribution of resources amongst the collective's members, as well as the ethical principles upon which the collective is maintained. In all these aspects, planning has a key role to play: it can orchestrate novel and innovative alignments of entities in its plans and decision-making processes; it can work to offset the power relations that generate unequal distributions of resources; and it can elaborate ethical principles for the governance of ecological space. However, ecological planning implies a new form of planning, one that is amenable to multiplicities of various kinds and one that is prepared to entertain future trajectories of development, which are somehow open and somehow disordered.

In the third case-study chapter, we looked in some detail at how networks operate to orchestrate spatial arrangements in practice (rather than just in the theoretical imagination). We compared two networks and their spaces: a 'fast' network of prescription and standardization and a 'slow' network of fluidity and diversity. We showed that the first of these aimed to establish a simplified ecology in which relational alignments allowed resources (for example, food, technology, customers) to be gathered in, while troublesome elements (for example, pollution, obesity, waste) were sifted out. This contrasted with the second network, which sought to bring as many aspects as possible within the ambit of its operations (for example, taste, culture, ecology, social relations) on the understanding that this would lead to unsavoury aspects being reduced in number (for example, pollutants, adverse health effects, cultural erosion). It was suggested that the second of these networks – which broadly conformed to Probynn's (2000: 57) conception of a 'palatable recombination of affect, eating and ethics' – might deliver some important insights for (ecological) strategies that aim to combine ('globalized') spatial relations and ('localized') spatial locations. In Massey's (2004) terms, it shows how the 'local' can be mobilized within 'global' networks so as to challenge dominant or Panoptic modes of spatialization.

In a variety of ways, then, the case-study chapters take forward the parallels between post-structuralism and ecology that were referenced in the opening chapter. Yet, while many commonalities between the two approaches have been identified (that is, their shared concern for relations and relationalism), there is perhaps a need now to clarify the benefits that might accrue to geography from any closer alignment of post-structuralist and ecological theories. In other words, what insights can be generated by a post-structuralist standpoint on ecological relations and what assistance might these insights provide to the practice of human geography? Some provisional answers to these questions have been given in the preceding chapters where it has been shown that a more ecologically sensitive form of post-structuralism can bring the range of relationships between social actors and material spaces to the fore in spatial analysis. The aim of this chapter is to provide a general context for this finding by showing how post-structuralist theory incorporates some measure of ecological thinking. The account provided below follows closely that given by Conley (1997), where it is argued that ecology lies close to the heart of post-structuralism. However, while this general insight is endorsed here, it is also proposed that post-structuralism should be tailored to the requirement for ecological action on the part of *humans*. Thus, it is suggested that post-structuralism may need to qualify its traditional anti-*humanism*, if it is to generate political progress on environmental issues. In other words, the chapter argues that post-structuralist geography can retain its radical or critical 'edge', if it sets humans *within* ecological relations but resists dissolving them into such relations.

Post-structuralism and ecology

As we saw at the beginning of Chapter 1, structuralism 'decentres' the human subject, so that change and development come to be seen as the unfolding of impersonal, underlying structures. One of the arenas in which Lévi-Strauss developed this structuralist approach was in his analysis of 'myth'. Lévi-Strauss argued that all forms of cultural organization are built upon myths of various kinds (Lévi-Strauss, 1964). Myths play a role in binding socio-cultural phenomena together and make patterns of social organization intelligible and meaningful. One myth of particular interest to Lévi-Strauss was that of the rational, male subject, controlling and dominating the human and nonhuman environment. According to Conley (1997: 43), Lévi-Strauss was concerned to show that these all-powerful male subjects are nothing but a 'living species'; thus, he argued, they are enmeshed in nature as well as culture ('nature is in and of culture', Conley, 1997: 51). Nevertheless, while Lévi-Strauss recognized that 'man' is embedded in nature, he also bemoaned modifications of nature by this same 'man'. In particular, he was concerned about the destruction of non-Western cultures and environments by the expansion of advanced industrial societies. He saw the spread of Western culture and associated economic practices as undermining the viability of non-Western cultural forms. This was to be deplored, he argued, because non-Western cultures tend to nurture their environments rather better than cultures in the West.

Interestingly, Conley suggests that, for Lévi-Strauss, respect for other (non-Western) peoples was an 'ethical' question:

> Upon humans as ethical beings devolves the responsibility to assume the duty of safe-guarding the diversity of the living. Unlike grasshoppers, tent caterpillars, locusts, spruce budworms or other causes of natural plagues, humans are aware of their actions. The biological human being needs to draw sustenance from animals and plants. Ethical human beings develop a sense of respect toward themselves, their kindred and other

species. Lévi-Strauss's ethics are set in a chain of creation in which all links are connected. An extinction of one species unfastens the whole, depriving other species of their right to live. (1997: 47)

Thus, while humankind in general must carry responsibility for its ecological actions, it is one particular version of humankind that must carry most responsibility as it tends to engender the most destruction: the Western self and its associated ego; 'the self and personal identity are a trap that cuts humanity off from the world' (Conley, 1997: 52). Lévi-Strauss therefore aligned his affirmation of human ethical duties towards the environment with his structuralist anti-humanism: humans can only act well towards nature once they realize that they are part and parcel of nature, thereby accepting a more humble role in the making and shaping of the world.

Lévi-Strauss's studies indicate that structuralism might be effectively associated with ecology. As we move towards post-structuralism the linkage between the two becomes even more pronounced. However, ecological post-structuralists initially pay rather more attention to the systemic qualities of eco-systems than to the specific duties of humans. One of the leading theorists in this regard is Michel Serres. As we have already seen (in Chapter 4), Serres is interested in relations, in the 'signals' or 'messages' that circulate across space through relational webs. Like Lévi-Strauss, he decentres the human from its previously privileged place in the firmament: 'man' is now part of a 'living organism'; he is made up of 'interlocking information or language systems, the most complex of which is biological language itself, that which orders all social organisations in humans' (Conley, 1997: 61). In Serres's view, these systems are turbulent and unstable, given to fluctuating movements. And humans too are subject to recurrent destabilizations: 'human bodies are in constant flow, maintaining a delicate balance between stasis, redundancy and disorder in themselves, among each other, and with the environment' (Conley, 1997: 62). Thus, Serres downplays the 'humanness' of humans: he shows that they are 'structurally similar to *all* of creation, both organic and inorganic' (Conley, 1997: 61). Humans are defined only by their embeddedness in (ecological) systems: in this context 'being – no longer separable from information – cannot define itself against another being, or an object. It is a complexity, both microcosm and macrocosm, part of a larger microcosm and macrocosm' (Conley, 1997: 64).

Again, however, knowing humans are addressed in Serres's theory, although these are not disembodied, disengaged humans; rather, they are humans as part of wider collectives, situated within ecological formations. Serres (1995) argues that, in the wake of changes in the global environment, humans must forge a 'natural contract' with the earth in order to establish 'balance' and 'reciprocity' with nature. This contract would explicitly aim to democratize (or 'horizontalize', Serres, 1995) the position that humans hold in the overall scheme of things. Democratization would bring humans literally 'down to earth'; they would be forced to realize that efforts at 'mastery' and control are doomed to failure; now only an engagement with unpredictability and disorder will suffice. Thus, for Serres, human actions in the domain of ecology must aim at the retention and promotion of biological diversity, but they must also be attuned to a context in which nature is defined by its incorporation in dynamic and complex systems. The 'politics of nature' must, therefore, aim to establish some sustainable mixture of order and disorder, chaos and calm, information and noise.

Further clues to the composition of a new post-structuralist politics are provided by Deleuze's collaborator Felix Guattari (2000) in his short book, *The Three Ecologies*. Like Serres, Guattari sees humans as located in a complex system that is constantly changing. Thus, Guattari devises a form of eco-subjectivity 'that is immanent and in constant becoming' (Conley, 1997: 93). This new eco-subjectivity unfolds in a territory of multiplicities and emergent relations; thus Guattari argues for a 'mobile subject', one that is 'affectively' engaged to ecological territory: 'beings, neither quite autonomous nor endowed with an immutable foundation, assembling for affective reasons on a common '*Grund*', an existential territory in movement and transformation, open onto becoming and process' (Conley, 1997: 94). These beings are bound to territory by relations of various kinds: 'humans interact with each other and the planet' (1997: 94).

Guattari here appears to take up Serres's systemic approach to ecological process and politics. However, he goes on to identify a form of politics that is overtly *ethical* in character. In Guattari's view, ecological action would challenge dominant ways of thinking that subordinate ecological entities to the working of capitalist economies. In Deleuzian/Guattarian terms, ecological politics should engender a *de*territorialization – that is, a 'flight' from dominant ways of thinking and restrictive forms of behaviour, so as to allow a reterritorialization – that is, an affective and relational engagement with ecological space. In order to achieve this double movement (de/reterritorialization) Guattari argues for a 'reconstruction of subjectivities' (Conley, 1997: 96). This consists of three aspects:

1. *Mental ecology*, including 'myriad relations from which we make selections and draw diagrams that contribute to the construction of ever-changing ecosystems' (Conley, 1997: 96). Mental ecologies are important because they not only shape perceptions of nature but also influence actions towards natural entities. As Conley (1997: 98) puts it: 'mental ecology consists of multiple relations in and with the world. By deterritorialising and reterritorialising, the subjects break off from a territory and build new, virtual worlds with the imaginative wherewithal that an ecological mode of thinking is best able to provide'. Through the imaginative generation of these new, virtual worlds, 'we trace new lines of flight, new diagrams [...] everything evolves in continually changing assemblages' (1997: 99).

2. *Social ecology*, including everyday practices of citizenship, consists of 'developing specific practices that will modify and reinvent the ways in which we live as couples or in the family, in an urban context or at work, etc'.

(Guattari, 2000: 34). Social ecology effectively means re-establishing the social bond, but in ways that are sensitive to ecological requirements.

3. *Machinic ecology* stimulates a reconsideration of nature. No longer is environmental action predicated on the simple defence of nature; now a more dynamic, evolutionary approach is required. This approach would be concerned not with discrete natural entities but with the complex assemblages in which these entities are inevitably situated. Thus, Guattari (2000: 66) suggests 'we might just name environmental ecology *machinic* ecology'.

These three aspects of eco-subjectivity give us some insight into a post-structuralist ethos for environmental action. We can see that such action must be conducted in three registers simultaneously: in the arena of concepts and visualizations (as in Chapter 6 above on planning); in the arena of social relations and political mobilization (as in Chapter 7 above on food); and in the arena of environmental action and the harnessing of dynamic ecological processes (as in Chapter 5 above on urban–rural distinctions). Guattari's eco-subject must also develop an acute spatial sensibility through these three registers. This sensibility must be strongly relational and strongly affectual; it must aim not at the controlling or closure of space, but rather at the artful steering of dynamic socio-spatial processes:

> By definition, the 'art of the eco' is process itself. A practice based on openness constitutes the very essence of an art of the science of ecology that goes through all existing ways of domesticating existential territories, modes of being, the body, the environment, the contextual assemblages of ethnic groups, including general rights of humanity. Vertical hierarchical power assemblages (*pouvoir*) are replaced by horizontal, spatial assemblages (*puissances*) that enable social change. (Conley, 1997: 103)

Guattari gives us, then, a forceful characterization of the eco-subject, the post-structuralist political ecologist working in new ways to challenge the ecologically damaging trajectory of contemporary capitalism. However, Guattari himself admits that his concern for 'subjectivity' may strike some (post-structuralist) readers as rather odd. As he says:

> in the name of the primacy of infrastructures, of structures or systems, subjectivity still gets a bad press, and those who deal with it, in practice or theory, will generally only approach it at arms length, with infinite precautions, taking care never to move too far away from pseudo-scientific paradigms, preferably borrowed from the hard sciences: thermodynamics, topology, information theory, systems theory, linguistics etc. It is as though a scientistic superego demands that psychic entities are reified and insists that they are only understood by means of extrinsic coordinates. Under such conditions, it is no surprise that the human and social sciences have condemned themselves to missing the intrinsically progressive, creative and auto-positioning dimensions of processes of subjectification. (2000: 36)

What is striking is that Guattari could be referring here to almost any aspect of the post-structuralist literature in making his complaint about the failures of the human and social sciences. As we have seen in previous chapters, topology, systems theory, information theory, linguistics have all fed into post-structuralism in one way or another, and all have generated a great deal of valuable work on relationality and the composition of space. However, as Guattari indicates, this literature has also struggled with subjectivity and in arguing for his 'three ecologies' he feels the need to reinstate the notion of the (eco-)subject. Thus, the question is raised as to whether this reinstatement moves us out of post-structuralism's traditional anti-humanism back into a humanistic frame of reference.

Relationality and reflexivity

The various post-structuralist contributions to ecological thinking presented above all propose forms of relational thinking as the most appropriate way to capture ecosystem dynamics. Claude Lévi-Strauss stresses the way culture is embedded in nature; Michel Serres sees nature as but one part of dynamic and turbulent systems, in which various entities are thrown together in unexpected and unpredictable ways; Felix Guattari describes nature in terms of territories of emergence and becoming, in which multiple processes flow both together and apart, thereby generating further rounds of complexity. Interestingly, despite their post-structuralist predilections, all these authors retain a concern for human actions and knowledges (especially Lévi-Strauss and Guattari). Their theorizing is aimed at generating some form of ecological action on the part of human actors and human social groupings. This is taken furthest by Guattari, when he calls for new forms of 'eco-subjectivity' based on revised mental, social and environmental sensibilities. Here, then, post-structuralism displays an avowed political–ecological intent.

The conjoined emphasis on relationalism and subjectivity means that these post-structuralist accounts emphasize the building of new connections between social and natural entities. They take an almost holistic approach to this endeavour and stress the way humans are necessarily encompassed within multiple sets of relations and multiple forms of belonging. In fact, Guattari goes so far as to argue that the 'human' is disintegrated into these relations and belongings. He says:

> 'rather than speak of the "subject", we should perhaps speak of components of subjec-
> tification, each working more or less on its own. This would lead us, necessarily, to
> re-examine the relation between concepts of the individual and subjectivity, and, above
> all, to make a clear distinction between the two. Vectors of subjectification do not
> necessarily pass through the individual, which in reality appears to be something like a
> 'terminal' for processes that involve human groups, socio-economic ensembles, data-
> processing machines, etc. Therefore, interiority establishes itself at the crossroads of
> multiple components, each relatively autonomous in relation to the other, and, if need
> be, in open conflict. (2000: 36)

What emerges here, then, is the notion of the 'relational subject', a form of identification or a locus of action that is 'made up' of many cross-cutting processes of 'subjectification'. This relational subject can be seen as an ecological subject – indeed, Guattari aligns the two in the notion of eco-subjectivity so that ecological action comes to be seen as a (key) form of relational action.

However, if we think back for a moment to Chapter 7, it will be recalled that in the analysis of ecological action in the food sector, relationality was conjured up in concert with 'reflexivity' (these two aspects were spliced together in the rather clumsy phrase 'relational reflexivity'). Thus, eco-subjectivity might be seen to have a dualistic quality. On the one hand, ecological action requires the establishment of new connections between subjects and objects so that ecological alignments (or 'partnerships') can be consolidated. On the other hand, eco-subjects must assess social and environmental relations in reflexive terms. Such reflexivity requires that a 'critical distance' is established between the subject and object so that the most appropriate course of (ecological) action can be ascertained. By combining these two aspects, we can suggest that the relational ethic provokes eco-subjects into an awareness of themselves as reflexive and knowing participants embedded within complex ecologies. We therefore arrive at a position where humans are seen as enmeshed within heterogeneous relations but also that they retain distinctive qualities as participants in such relations. Thus, while we no longer see humans as disembodied subjects, or as actors who always and everywhere retain a privileged status, we nevertheless recognize that humans hold reflexive capacities that set them apart in some way from other entities.

In identifying how we might understand the role of different entities in relational contexts, we can turn to Ian Hacking's (1999a) attempt to redraw the rather crude distinction that currently exists (in geography, as elsewhere) between 'nature' and 'society'. In so doing, he introduces the notion of different 'kinds' so as to focus our attention on differing forms of socio-ecological action. In particular, Hacking introduces a distinction between 'interactive' and 'indifferent' kinds. In his view, humans are interactive kinds because they can reflect upon their incorporation into socio-material relationships and can act upon these reflections. In particular, humans can use language-based resources to assess how they are being represented (by, for instance, other humans) or how they are being acted upon (by, for instance, heterogeneous sets of relations). Hacking uses the term 'interactive kinds' to show how forms of agency are linked to the ways in which people conceptualize themselves and how they then act upon these conceptualizations. He claims that other entities do not behave in quite this way. In illustrating this point, Hacking cites (following Pickering, 1994) the example of quarks, and he argues that, although they are quite capable of action, quarks are not aware of the classifications made about them: 'Our knowledge about quarks affects quarks, but not because they become aware of what we know, and act accordingly'. Hacking calls entities of this type 'indifferent kinds': 'the classification "quark" is indifferent in the sense that calling a quark a quark makes no difference to the quark' (1999a: 105).

In this account, we still discern the effect of complex, heterogeneous relations on the constitution of particular entities (Hacking concedes that being 'interactive' is dependent upon immersion in such relations); yet, it is proposed that natural and social entities will respond in different ways to their positioning within particular relational arrangements and these differences are attributable to some stable and immutable characteristics that are not fully reducible to surrounding relations (in other words, there is a humanist residue, a trace of subjectivity). These differences hinge on the reflective abilities of humans, abilities that derive from social relations (in particular, shared languages and cultures). Therefore, (human) entities cannot always simply be thought of as potential 'allies', to be enrolled in processes of relational fabrication (as argued by Latour, 1987, for instance), for they can make conscious, reflexive responses to the act of enrollment and can thereby alter the whole functioning of the relational configuration. (Although other entities can obviously modify their various associations these modifications are not normally based on reflexive processes of deliberation).

Hacking illustrates how the distinction between interactive and indifferent kinds can be brought to bear in his book, *Mad Travellers* (1999b), which deals with the appearance and disappearance of a mental illness known as 'fugue'. The term 'fugue' referred to a strange compulsion to wander, a compulsion that was preceded by insomnia, migraine and amnesia. It was initially diagnosed in 1887, but it remained a recognized medical condition for only twenty years. Hacking thus calls fugue a 'transient mental illness': it is a social phenomenon that emerges from a particular set of 'ecological conditions'; once these conditions changed then the phenomenon disappeared. In this case, the ecology that allowed the illness to flourish included the following: systems of detection, notably identity-card checks on travellers; a taxonomy that recognized certain behaviours as illnesses; a cultural polarity that valorized certain forms of behaviour and disapproved of others; and the apparent need on the part of a number of individuals to engage in behaviours commensurate with the condition. This last aspect draws our attention to the interactive nature of transient mental illnesses: Hacking explains that, during the early stages of fugue development, patients and doctors together elaborated a set of symptoms that came to distinguish the illness. Hacking emphasizes the interaction of doctor and patient and explains that each was very accommodating to the expectations of the other. In the process of interaction the condition known as 'fugue' began to take shape such that it came to be seen as a discrete phenomenon.

This account of 'fugue' shows how the illness was nested in a complex ecology. But what made this a *transient* mental illness was its reliance on the *social* aspects of this ecology, and when those aspects changed so did the illness. The ecology of fugue can be compared to the conditions that surround a *non-transient* mental illness, for instance, schizophrenia. Here we find a condition that appears not only as a result of social causes but also *physical* factors such as genes which stimulate the onset of the disease (Crichton, 2000). The ecology of schizophrenia has thus come to be understood as a mixture of both interactive and indifferent kinds. While the causes of fugue can be investigated through the analysis of systems of classification and doctor/patient relations, understanding the causes of schizophrenia requires some attention to the interaction between natural and social entities.

Hacking emphasizes that processes of ecological symbiosis involve entities of different types, and in order to distinguish these types he reinstates a division between humans and nonhumans in the distinction between interactive and indifferent kinds. In comparing fugue and schizophrenia, he draws attention to the central role of reflective action in the former and the diminished significance of such action in the latter. Thus, Hacking asserts a need to attend to the particular forms of reflexive calculation that are associated with human behaviour. However, he still emphasizes that human behaviour is embedded within complex ecologies. Hacking also believes that ecologies are heterogeneously composed. But he emphasizes that we can only make sense of ecology-dependent action if we retain a fundamental distinction: humans are (often?) 'interactive' and (most?) nonhumans are 'indifferent'. This distinction is fundamental because it remains potentially salient even when set within heterogeneous sets of (ecological) relations. It implies that different entities retain the potential for differing behaviours, despite the precise configuration of any particular ecology. Ecological action therefore needs to be attentive to the 'mix' of entities so that ecological strategies are tailored appropriately, that is, where ecological conditions stem from the actions of interactive kinds, a rather different approach is required to conditions that depend on indifferent kinds. Thus, the components of subjectivity identified by Guattari will vary according to the ecological contexts in which subjects emerge.

Eco-subjectivity and spatial strategy

Hacking's ecological approach allows us to accompany post-structuralism into relational space, so that we can describe the heterogeneous relations that comprise complex ecosystems. At the same time, however, it insists we take note of a fundamental distinction between natural and social actors, one that is based upon their differing abilities to reflect upon, and thus change, the social arrangements in which they are enmeshed. As Hacking (1999a: 32) says: 'people are aware of what is said about them, thought about them, done to them', and they act on the basis of such awareness. And, as this awareness extends to what is done to others, including nonhumans, it provides a moral and ethical dimension to human action (Hacking, 1999c: 13). For Hacking, people have the potential to become moral agents – morality is 'firmly rooted in human values and the potential for self-awareness' (Hacking, 1999a: 59) – and this is not something that applies to indifferent kinds.

It is no surprise that some of those most concerned with the pursuit of an ecological ethics should concur that humans, for all their embeddedness in the complex relations of 'natureculture' (Haraway, 1997), continue to carry responsibility for the fate of nonhumans. As Kate Soper argues, there can be no ethical prescription that does not presuppose some kind of demarcation between humans and nature:

> Unless human beings are differentiated from other organic and inorganic forms of being, they can be made no more liable for the effects of their occupancy of the ecosystem than can any other species, and it would make no more sense to call upon them to desist from destroying nature than to call upon cats to stop killing birds. (1995: 160)

In other words, the need to act 'ecologically' is a human need, one that is given voice within human languages and cultures. However, following the insights of post-structuralist geography, we need to consider how human relations are woven into heterogeneous ecologies. By attending to the (spatial) zone where nature and society 'meet', we might begin to elaborate an ecological approach that displays the full ecological consequences of human action. It may also enable us to situate the components of subjectification (identified by Guattari) in spaces of multiplicity and affect (McCormack, 2003). This approach perhaps give rise to a form of 'relational ethics', one which emphasizes 'the situatedness of ethical agency and the extralinguistic connectivities of the ethical community' (Whatmore, 1997: 44). Such an ethics will require attention to the heterogeneous composition of human action, the nonhumans that lend themselves to this action and the ecosystem in which it unfolds. It also implies a very human sense of responsibility towards both nonhumans and ecosystems as subjects are composed from relations that extend into ecological contexts (Murdoch, 2001). In such circumstances it seems obviously beneficial for humans to be 'extended into' rich and diverse, as opposed to simple and denuded, ecological surroundings.

The relational ethic described by Whatmore (1997) can be seen not only as an 'ecological ethic' (Conley, 1997) but also more generally as a 'spatial ethic'. In previous chapters, it has been shown that space is relational in nature and that spatial 'permanences' (to return to Harvey's, 1996, term) are carved out of complex and dynamic processes of change. The turn to more overtly ethical questions leads us to consider the kinds of permanences that should be provided and supported. The principles of ecology are of some help in providing an answer as they propose that permanences should consist of alignments or partnerships between natural and social entities (Merchant, 2003). This brings us back to Latour's (2004) proposals for political ecology outlined in Chapter 6. Latour argues that the aim of political ecology is not to root politics in nature; rather it is to 'convoke a single collective' (2004: 29) made up of 'associations of humans and non-humans', associations in which humans and nonhumans 'exchange properties' (2004: 61). All that matters, in this approach 'is the production of a common world, one that [...] is offered to the rest of the collective as an occasion to unite' (2004: 141). Permanences should therefore aim to embrace a range of differing entities on terms that sustain the well-being of all.

Geography clearly has a key role to play in the building of Latourian 'common worlds'. As we saw in Chapter 1, Richard Peet (1998: 1) sees the synthetic core of geography as lying in the study of nature–society relations: to repeat, 'geography looks at how society shapes, alters and increasingly transforms the natural environment, creating humanised forms from stretches of pristine nature, and then sedimenting layers of socialisation, one within the other, one on top of the other, until a complex natural-social landscape results'. Given this focus, clearly geography should be able to contribute its extensive reserves of knowledge to new processes of ecological 'world-building'. And yet, there is some doubt about geography's abilities in this regard. For instance, Noel Castree (2003: 207) observes that 'it is a peculiar fact that a discipline [geography] which, in part, defines itself as the study of society–environment relations has conspicuously failed to engage with questions on the political status of the non-human'. Thus, Castree goes on to suggest that geography needs to:

- Abandon the idea that political rights and entitlements only apply to people.
- Confront the problem of defining political subjects in a world where the boundaries between humans and nonhumans are hard to discern.
- Expand political reasoning to include nonhumans 'without resorting to the idea that the latter exist "in themselves"' (Castree, 2003: 208).

Castree encourages geographical work that thinks through the significance of the 'relational turn' in order to develop a new geographical vocabulary. This vocabulary should be capable of describing and assessing the heterogeneous complexities that now animate relational spaces. However, he also emphasizes that this vocabulary must be accompanied by 'substantive political concepts that ground new forms of practice (2003: 208). This brings us back once again to the 'reflexive subject'. It has been suggested above that a geographical engagement with political ecology must be predicated on the assertion of new forms of 'eco-subjectivity'. If we return to Guattari's description of the 'three ecologies', we can perhaps see a little more clearly how geographical subjectivities might be re-composed:

1. *Mental ecology*, which would include the relationship between geography as an intellectual discipline and the 'external' geographical world. Geography plays an important role in 'performing' the world, of bringing it into being through representational and non-representational practices. In the new political–ecological context, geography needs to ensure it plays this role in ways that enable the building of new, virtual worlds which 'trace new lines of flight, new diagrams' (Conley, 1997: 99). As previous chapters have argued, these new diagrams will need to sketch out some alignment between topographical and topological spaces – that is, between spatial locations and spatial relations, in ways that bolster ecological integrity.
2. *Social ecology* consists of developing specific practices that will help ecologically sensitive social formations come into being. This is perhaps the task

with which *human* geography is most familiar. As shown in Chapter 1, post-structuralist geography has spent a great deal of time looking at social inclusions and exclusions. It has acted to open out the geographical enterprise so that it can embrace previously excluded groups and identities. However, this concern for 'otherness' and 'marginality' might be turned more explicitly towards a concern for nonhuman 'others', to those natural entities that have to yet to be brought within social collectives.

3. *Machinic ecology* specifies that any incorporation of nonhumans into the geographical collective should be predicated not upon the simple defence of discrete entities and their associated spaces but on a concern for dynamic and complex systems of heterogeneous relations. As Guattari (2000: 66) puts it: 'natural equilibriums will be increasingly reliant upon human intervention, and a time will come when vast programmes will need to be set up in order to regulate the relationship between oxygen, ozone, and carbon dioxide'. Geography can clearly play a key role in articulating such programmes.

These three aspects of 'eco-subjectivity' help to define geographical subjectivity a little more closely. They suggest that the relational perspective now pre-eminent within human geography must be thought in the three registers simultaneously so that spatial imaginaries ('mental ecologies') are aligned with social practices ('social ecologies') and an assessment of general ecological effects ('machinic ecologies'). The discipline of geography is therefore being asked to reflexively assess how it might generate new and innovative relations between itself and the ecological world. New geo-subjectivities are proposed that embrace the mental, social, machinic ecologies identified by Guattari. While differing 'geo-' or 'eco-' subjects will interiorize these ecologies in differing ways, all will maintain an acute sensitivity to interactions between societies and natures, humans and nonhumans, territories and relations, singularities and multiplicities, orders and disorders. These and other such (ecological) interactions define the spatial imagination of a post-structuralist, 'more-than-human' geography (Whatmore, 1999).

Conclusion

In this chapter, an ecological perspective on post-structuralism has been outlined in order to show that post-structuralist geography might best be positioned at the interfaces between nature and society and between human and nonhuman worlds. The suggestion has been made here that what defines geography is exactly this focus on natural and social relations. It has been claimed that 'heterogeneity', the mingling of various entities in complex assemblages, networks and/or systems, might now comprise geography's main intellectual concern. This conclusion has been reached not primarily because the ecological crisis so obviously constitutes humankind's greatest challenge, but because the distinctive nature of the geographical enterprise can be discerned most clearly at the point where the 'social' becomes embedded in the 'natural' (or where the 'human' becomes immersed in the 'nonhuman'). Geography becomes, then, the study of relations, it investigates the various ways in which entities of differing kinds are connected and disconnected. But more than this, it shows that the entities themselves are relationally composed so that any coherence they achieve is only provisional and reversible, something that is carved out of dynamic, unstable, turbulent contexts and something which always threatens to dissipate into such contexts.

While the relational perspective has been largely endorsed in the preceding pages, this final chapter has added one or two qualifications to the overall analysis. Yes, entities may be relational achievements, but the 'centering' of relations in subject positions can lend entities a stability that begins to look like a clear distinction between the entity and the relation. In actual fact, of course, this distinction emerges so frequently it gets given many names – organic/ inorganic, human/nonhuman, social/natural. In the preceding discussion we added another distinction into the mix: interactional/indifferent. This suggests that some entities (usually, but not always, humans) acquire the ability to reflect upon the relations that comprise or surround them. Through processes of reflection, bodies are made to move, relations are made to change, and new classifications are made to come into existence. Given the significance of reflexive action, it has been suggested that modes of subjectivity might be thought of as 'reflexive relationalities' (or perhaps 'relational reflexivities'), so that reflections *upon* action can never be fully distinguished from the heterogeneous relationships that *facilitate* action.

Moreover, it has been argued that the modes of subjectification performed within geography should be oriented to ecological relationalities – that is, to the promotion of human–nonhuman partnerships that work to sustain biodiversity and other such ecological 'goods'. In this context, geography obviously has an important role to play: it can provide ways of analysing, understanding and promoting ecological ways of being and it can be attentive to the shifts in social and spatial arrangements that will be required if such ways of being are to be established in practice. Geography thus potentially lies at the heart of processes of 'eco-subjectification' for it can help to build alignments between the mental, social and machinic ecologies that Guattari and others see as so significant at the present time.

In conclusion, then, we can suggest that post-structuralism in geography is not simply a theoretical endeavour. It is a way of shifting spatial imaginaries so that new forms of geographical practice come into being. From a post-structuralist perspective, no longer should geographical practitioners be detached from heterogeneity; like the planners examined in Chapter 6, they should be subsumed within complexities and multiplicities of various kinds. Yet, the imperative here is not simply 'subsumption for its own sake': it is 'subsumption with a purpose' and the purpose is a strengthening of heterogeneous associations within given ecological contexts. Thus, the aim of geographical practice becomes not some form of detached spatial 'mastery' but rather the iterative development of ecological 'steering mechanisms'. These mechanisms must necessarily be sensitive to interactions between natures and societies, humans and nonhumans, knowledges and materials, singularities and multiplicities, territories and relations. They must also comprise effective interventions in processes of spatial (de-)formation so that stronger alignments between all the interacting phenomena are established (in line with ecological principles).

'Steering the spatial' is perhaps not a slogan likely to inspire great enthusiasm, but it seems well-suited to an era in which complex socio-natural processes *always* escape geography's dominant modes of ordering. In this context, the value of post-structuralism is its simultaneous attention to processes of ordering *and* disordering and it has been argued that post-structuralism's demand that both sets of processes be integrated into the *same* spatial framework provides a useful starting point for geographical analysis. In the preceding pages this framework has been identified and investigated and it has been suggested that it be used to assist the efforts of political-ecologists, planners, food movements and all those various others who now strive to bring rich and diverse ecologies into being. In other words, geo-subjectivity should now become a core component of eco-subjectivity so that heterogeneous and relational spatialities are consolidated in both theory and practice.

SUMMARY

In this chapter, the parallels between post-structuralist theory and ecological thought have been identified and discussed. It was argued that a number of post-structuralist authors, notably Michel Serres and Felix Guattari have explictly addressed ecological concerns in their works. Both these theorists believe social formations should be seen as set within complex and dynamic ecological systems. They therefore emphasize the turbulent character of nature–society relations. However, both also recognize that social formations (especially in the capitalist West) are threatening nature as never before. Thus, Guattari calls for the assertion of new modes of 'eco-subjectivity'. Drawing upon Hacking's work, it was suggested that 'eco-subjectivity' can be thought of in both relational and reflexive terms: it requires human subjects to acknowledge their embeddedness in ecological formations while also requiring that they consider the most appropriate forms of ecological action. This notion of relational-reflexive eco-subjectivity, it was argued, provides a model for geographical practice, so that new modes of geo-subjectivity might come into being which aim to align geography more closely to ecology.

FURTHER READING

The key text on relationships between post-structuralism and ecology is Vera Andermatt Conley's (1997) book, *Ecopolitics: The Environment in Poststructuralist Thought*. On relationality and reflexivity, Ian Hacking's (1999) book, *The Social Construction of What?*, ranges widely but is very accessible. Again, Bruno Latour's (2004) *Politics of Nature* has general relevance to the ideas expressed above.

References

Abercrombie, P. (1933) *Town and Country Planning*. London: Oxford University Press.

Allen, J. (1999) 'Afterwords: open geographies', in D. Massey, J. Allen, and P. Sarre (eds), *Human Geography Today*. Cambridge: Polity Press. pp. 323–8.

Allen, J. (2003) *Lost Geographies of Power*. Oxford: Blackwell.

Allen, J., Massey, D. and Cochrane, A. (1998) *Rethinking the Region*. London: Routledge.

Amin, A. (2002) 'Spatialities of globalisation', *Environment and Planning A*, 34: 385–99.

Amin, A. and Graham, S. (1997) 'The ordinary city', *Transactions of the Institute of British Geographers*, 22: 411–29.

Ansell-Pearson, K. (2002) *Philosophy and the Adventure of the Virtual: Bergson and the Time of Life*. London: Routledge.

Barker, P. (1998) *Michel Foucault: An Introduction*. Edinburgh: Edinburgh University Press.

Barnes, T. and Duncan, J. (1991) *Writing Worlds: Discourse, Text and Metaphor in the Representation of Landscape*. London: Routledge.

Barnes, B., Bloor, D. and Henry, J. (1996) *Scientific Knowledge: A Sociological Analysis*. London: Athlone Press.

Barry, A. (2001) *Political Machines: Governing a Technological Society*. London: Athlone Press.

Barthes, R. (1975) *S/Z*. New York, NY: Hill and Wang.

Barthes, R. (1993) *Mythologies*. London: Penguin.

Batty, M. (1985) 'Formal reasoning in urban planning', in M. Breheny and A. Hooper (eds), *Rationality in Planning: Critical Essays on the Role of Rationality in Urban and Regional Planning*. London: Pion. pp. 98–117

Beardsworth, A. and Keil, T. (1997) *Sociology on the Menu*. London: Routledge.

Beck, U. (1992) *Risk Society. Towards a New Modernity*. London: Sage Publications.

Bell, D. and Valentine, G. (eds) (1995) *Mapping Desire: Geographies of Sexuality*. London: Routledge.

Bell, D. and Valentine, G. (1997) *Consuming Geographies: We Are Where We Eat*. London: Routledge.

Belsey, C. (2002) *Poststructuralism: A Very Short Introduction*. Oxford: Oxford University Press.

Bhabba, H. (1994) *The Location of Culture*. London: Routledge.

Bingham, N. and Thrift, N. (2000) 'Some instructions for travellers: the geography of Bruno Latour and Michel Serres', in M. Crang and N. Thrift (eds), *Thinking Space*. London: Routledge. pp. 281–301.

Binnie, J. (1997) 'Coming out of geography: towards a queer epistemology', *Environment and Planning D: Society and Space*, 15: 223–37.

Blowers, A. (1993) 'The time for change', in A. Blowers (ed.), *Planning for a Sustainable Environment: A Report by the Town and Country Planning Association*. London: Earthscan. pp. 1–31.

Bondi, L. (1990) 'Feminism, postmodernism and geography: space for women?', *Antipode*, 22: 156–67.

Bonta, M. and Protevi, J. (2004) *Deleuze and Geophilosophy: A Guide and Glossary*. Edinburgh: Edinburgh University Press.

Bowers, J. (1992) 'The politics of formalism', in M. Lea (ed.), *Contexts of Computer-mediated Communication*. London: Harvester Wheatsheaf. pp. 232–61.

Bowker, G. and Star, S.L. (2000) *Sorting Things Out: Classification and Practice*. Cambridge, MA: MIT Press.

Boym, C. (2001) 'My McDonald's'. *Gastronomica*, 1: 6–8.

Braudel, F. (1977) *Afterthoughts on Material Civilization*. Baltimore, MD: Johns Hopkins University Press.

Breheny, M. (1999) 'People, households and houses: the basis of the "great housing debate" in England', *Town Planning Review*, 70: 275–93.

Breheny, M. and Hooper, A. (eds) (1985) *Rationality in Planning: Critical Essays on the Role of Rationality in Urban and Regional Planning*. London: Pion.

Brown, S.D. (2002) 'Michel Serres: science, translation and the logic of the parasite', *Theory, Culture and Society*, 19: 1–27.

Bryson, J. and Crosby, B. (1992) *Leadership in the Common Good*. San Francisco, CA: Jossey Bass.

Butler, J. (1997) *The Psychic Life of Power: Theories in Subjection*. Stanford, CA: Stanford University Press.

Butler, J. (2004) 'Jacques Derrida', *London Review of Books*, 4 November, pp. 32–3.

Callon, M. (1986) 'Some elements in a sociology of translation', in J. Law (ed.), *Power, Action, Belief: A New Sociology of Knowledge*. London: Routledge and Kegan Paul. pp. 196–223.

Callon, M. (1991) 'Techno-economic networks and irreversibility', in J. Law (ed.), *A Sociology of Monsters: Essays on Power, Technology and Domination*. London: Routledge. pp. 132–61.

Callon, M. (1992) 'The dynamics of techno-economic networks', in R. Coombs, P. Saviotti and V. Walsh (eds), *Technological Change and Company Strategies: Economic and Sociological Perspectives*. London: Harcourt Brace Jovanovich. pp. 77–102.

Callon, M. and Latour, B. (1981) 'Unscrewing the big Leviathan: how actors macro-structure reality and how sociologists help them to do so', in K. Knorr-Cetina and A. Cicourel (eds), *Advances in Social Theory: Towards an Integration of Micro- and Macro-sociologies*. London: Routledge and Kegan Paul. pp. 277–303.

Callon, M. and Law, J. (1995) 'Agency and the hybrid collectif', *South Atlantic Quaterly*, 94: 481–507.

Callon, M. and Law, J. (1997) 'After the individual in society: lessons on collectivity from science, technology and society', *Canadian Journal of Sociology*, 22: 165–82.

Callon, M. and Law, J. (2004) 'Introduction: absence–presence, circulation, and encountering in complex space', *Environment and Planning D: Society and Space*, 22 (1): 3–11.

Callon, M., Law, J. and Rip, A. (1986) *Mapping the Dynamics of Science and Technology*. London: Macmillan.

Capatti, A. (1999) 'The traces left by time', *Slow*, 17: 4–6.

Castells, M. (1983) *The City and the Grassroots*. London: Edward Arnold.

Castree, N. (2003) 'Environmental issues: relational ontologies and hybrid politics', *Progress in Human Geography*, 27: 203–11.

Chadwick, G. (1971) *A Systems View of Planning: Towards a Theory of Urban and Regional Planning*. Oxford: Pergamon.

Cherry, G.E. and Rogers, A. (1996) *Rural Change and Planning*. London: Spon.

Clapson, M. (2000) 'The suburban aspiration in England since 1919', *Contemporary British History*, 14: 151–74.

Clifford, J. and Marcus, G. (eds) (1986) *Writing Culture: The Poetics and Politics of Ethnography*. Berkeley, CA: University of California Press.

Cloke, P., Philo, C. and Sadler, D. (1991) *Approaching Human Geography*. London: Chapman Publishing.

Cole, S. (1992) *Making Science: Between Nature and Society*. Cambridge, MA: Harvard University Press.

Conley, V.A. (1997) *Ecopolitics: The Environment in Poststructuralist Thought*. London: Routledge.

Connelly, W.E. (1999) *Why I Am Not a Secularist*. Minneapolis, MN: University of Minnesota Press.

Corbridge, S. (1993) 'Colonialism, post-colonialism and the political geography of the third world', in P. Taylor (ed.), *Political Geography of the Twentieth Century*. London: Belhaven. pp. 173–205.

Cosgrove, D. and Daniels, S. (eds) (1988) *The Iconography of Landscape*. Cambridge: Cambridge University Press.

Council for the Protection of Rural England (CPRE) (1996) *The cluttered countryside*. London: Council for the Protection of Rural England.

Crandell, G. (1993) *Nature Pictorialized: 'The View' in Landscape History*. Baltimore, MA: The Johns Hopkins University Press.

Crang, P. (1996) 'Displacement, consumption and identity', *Environment and Planning A*, 28: 47–67.

Crang, M. and Thrift, N. (2000) 'Introduction', in M. Crang and N. Thrift (eds), *Thinking Space*. London: Routledge. pp. 1–30.

Cresswell, T. (1996) *In Place/Out of Place: Geography, Ideology and Transgression*. Minneapolis, MN: University of Minnesota Press.

Cresswell, T. (2000) 'Falling down: resistance as diagnostic', in J. Sharp, P. Routledge, C. Philo and R. Paddison (eds), *Entanglements of Power: Geographies of Domination/ Resistance*. London: Routledge. pp. 256–68.

Crichton, P. (2000) 'The uses and abuses of schizophrenia', *Times Literary Supplement*, 31 March, pp. 14–15.

Critchley, S. (1996) 'Angel in disguise: Michel Serres's attempts to re-enchant the world', *Times Literary Supplement*, 19 January, pp. 3–4.

Critser, G. (2004) *Fatland: How Americans Became the Fattest People in the World*. London: Penguin.

Danaher, G., Schirato, T. and Webb, J. (2000) *Understanding Foucault*. London: Sage Publications.

Darier, E. (1999) 'Foucault and the environment', in E. Darier (ed.), *Discourses of the Environment*. London: Blackwell. pp. 1–34.

Davidson, A. (1994) 'Ethics and aesthetics: Foucault, the history of ethics and ancient thought', in G. Gutting (ed.), *The Cambridge Companion to Foucault*. Cambridge: Cambridge University Press. pp. 115–40.

de Bolla, P. (2004) 'In the butcher's shop: review of "Francis Bacon: the logic of sensation" by Gilles Deleuze', *London Review of Books*, 23 September, pp. 19–20.

de Certeau, M. (1984) *The Practice of Everyday Life*. Berkeley, CA: University of California Press.

de Saussure, F. (1986 [1916]) *Course in General Linguistics*. La Salle, IL: Open Court Publishing.

Dean, M. (1999) *Governmentality*. London: Sage Publications.

Dehaene, M. (2004) 'Urban lessons for the modern planner: Patrick Abercrombie and the study of urban development', *Town Planning Review*, 75: 1–30.

Delanda, M. (2002) *Intensive Science and Virtual Philosophy*. London: Continuum.

Deleuze, G. (1988) *Foucault*. Minneapolis, MN: University of Minnesota Press.

Deleuze, G. (2004) *Francis Bacon: The Logic of Sensation*. London: Continuum.

Deleuze, G. and Guattari, F. (1987) *A Thousand Plateaus: Capitalism and Schizophrenia*. Minneapolis, MN: University of Minnesota Press.

Denzin, N. and Lincoln, Y. (1994) 'Introduction: entering the field of qualitative research', in N. Denzin and Y. Lincoln (eds), *Handbook of Qualitative Research*. London: Sage Publications. pp. 1–17.

Derrida, J. (1998 [1967]) *Of Grammatology*. Baltimore, MD: The Johns Hopkins University Press.

Descola, P. (1992) 'Societies of nature and the nature of society', in A. Kuper (ed.), *Conceptualising Society*. London: Routledge. pp. 44–63.

Department of Environment, Transport and the Regions (DETR) (2000) *Planning Policy Guidance Note 3: Housing*. London: DETR.

Doel, M. (1999) *Poststructuralist Geographies: The Diabolical Art of Spatial Science*. Edinburgh: Edinburgh University Press.

Doel, M. (2000) 'Un-glunking geography: spatial science after Dr Seuss and Gilles Deleuze', in M. Crang and N. Thrift (eds), *Thinking Space*. London: Routledge. pp. 117–35.

Doel, M. (2004) 'Poststructuralist geographies: the essential selection', in P. Cloke, P. Crang and M. Goodwin (eds), *Envisioning Human Geographies*. London: Arnold. pp. 146–71.

Driver, F. (1994) 'Bodies in space: Foucault's account of disciplinary power', in C. Jones and R. Porter (eds), *Re-assessing Foucault*. London: Routledge. pp. 279–89.

Eder, K. (1993) *The New Politics of Class: Social Movements and Cultural Dynamics in Advanced Societies*. London: Sage Publications.

Eder, K. (1996) *The Social Construction of Nature: A Sociology of Ecological Enlightenment*. London: Sage Publications.

Eldon, S. (2001) *Mapping the Present: Heidegger, Foucault and the Project of Spatial History*. London: Continuum.

Environment Agency (2002) *The Urban Environment in England and Wales: A Detailed Assessment*. Bristol: Environment Agency.

Faludi, A. (1973) *Planning Theory*. Oxford: Pergamon.

Fantasia, R. (1995) 'Fast Food in France', *Theory and Society*, 24: 201–43.

Fiddes, N. (1992) *Meat: A Natural Symbol*. London: Routledge.

Fine, G.A. (1996) *Kitchens: The Culture of Restaurant Work*. Berkeley, CA: University of California Press.

FitzSimmons, M. (2004) 'Engaging ecologies', in P. Cloke, P. Crang and M. Goodwin (eds), *Envisioning Human Geography*. London: Arnold. pp. 30–47.

FitzSimmons, M. and Goodman, D. (1998) 'Incorporating nature: environmental narratives and the reproduction of food', in B. Braun and N. Castree (eds), *Remaking Reality: Nature at the Millennium*. London: Routledge. pp. 194–220.

Flynn, T. (1994) 'Foucault's mapping of history', in G. Gutting (ed.), *The Cambridge Companion to Foucault*. Cambridge: Cambridge University Press. pp. 28–46.

Flyvbjerg, B. (1998) *Rationality and Power: Democracy in Practice*. Chicago, IL: University of Chicago Press.

Flyvbjerg, B. and Richardson, T. (2002) 'Planning and Foucault: in search of the dark side of planning theory', in P. Allmendinger and M. Tewdwr-Jones (eds), *Planning Futures: New Directions for Planning Theory*. London: Routledge. pp. 44–61.

Forester, J. (1989) *Planning in the Face of Power*. Berkeley, CA: University of California Press.

Forester, J. (1999) *The Deliberative Practitioner: Encouraging Participatory Planning Processes*. Cambridge, MA: MIT Press.

Foucault, M. (1972) *The Archaeology of Knowledge*. London: Tavistock.

Foucault, M. (1978) *Madness and Civilisation*. London: Tavistock.

Foucault, M. (1979) *Discipline and Punish: The Birth of the Prison*. Harmondsworth: Penguin.

Foucault, M. (1980) *The Order of Things: An Archaeology of the Human Sciences*. London: Tavistock.

Foucault, M. (1981) *History of Sexuality*. Vol. 1, *An Introduction*. London: Penguin.

Foucault, M. (1982) 'The subject and power', in H. Dreyfus and P. Rabinow, *Michel Foucault: Beyond Structuralism and Hermeneutics*. Brighton: Harvester Press. pp. 208–27.

Foucault, M. (1986) 'Space, knowledge and power', in P. Rabinow (ed.), *The Foucault Reader*. London: Penguin. pp. 239–54.

Foucault, M. (1988) *History of Sexuality*. Vol. 2, *The Use of Pleasure*. London: Penguin.

Foucault, M. (1991) 'Governmentality', in C. Gordon (ed.), *The Foucault Effect: Studies in Governmentality*. London: Harvester Wheatsheaf. pp. 87–104.

Foucault, M. (2004) *Abnormal*. London: Verso.

Franklin, A. (1999) *Animals and Modernity*. London: Sage Publications.

Friedmann, J. (1987) *Planning in the Public Domain: From Knowledge to Action*. Princeton, NJ: Princeton University Press.

Fuller, S. (1994) 'Making agency count: a brief foray into the foundations of social theory', *American Behavioural Scientist*, 37: 741–53.

Gibson-Graham, J.K. (2000) 'Poststructural interventions', in E. Sheppard and T. Barnes (eds), *A Companion to Economic Geography*. Oxford: Blackwell. pp. 95–109.

Girardet, H. (1993) *The Gaia Atlas of World Cities*. New York, NY: Anchor Books.

Girardet, H. (1999) *Creating Sustainable Cities*. Dartington: Green Books.

Goodchild, P. (1996) *Deleuze and Guattari: An Introduction to the Politics of Desire*. London: Sage Publications.

Goodman, D. (1999) 'Agro-food studies in the "age of ecology": nature, corporeality, bio-politics', *Sociologia Ruralis*, 39: 17–38.

Goodman, D., Sorj, B. and Wilkinson, J. (1987) *From Farming to Biotechnology*. London: Routledge.

Gordon, C. (1991) 'Governmental rationality: an introduction', in C. Gordon (ed.), *The Foucault Effect: Studies in Governmentality*. London: Harvester Wheatsheaf. pp. 1–51.

Gottdiener, M. (1997) *The Theming of America*. Boulder, CO: Westview Press.

Graham, S. and Healey, P. (1999) 'Relational concepts of space and place: issues for planning theory and practice', *European Planning Studies*, 7: 623–46.

Green, B. (2002) 'The farmed landscape: the ecology and conservation of diversity', in J. Jenkins (ed.), *Remaking the Landscape*. London: Profile Books.

Gregory, D. (1994) *Geographical Imaginations*. Oxford: Blackwell.

Guattari, F. (2000) *The Three Ecologies*. London: Athlone Press.

Gutting, G. (1994) 'Michel Foucault: a user's manual', in G. Gutting (ed.), *The Cambridge Companion to Foucault*. Cambridge: Cambridge University Press. pp. 1–27.

Gutting, G. (2001) *French Philosophy in the Twentieth Century*. Cambridge: Cambridge University Press.

Hacking, I. (1983) *Representing and Intervening*. Cambridge: Cambridge University Press.

Hacking, I. (1986) 'Making up people', in T. Heller, M. Sosna and D. Wellberry (eds), *Reconstructing Individualism: Autonomy, Individuality, and the Self in Western Thought.* Stanford, CA: Stanford University Press. pp. 222–36.

Hacking, I. (1990) *The Taming of Chance.* Cambridge: Cambridge University Press.

Hacking, I. (1999a) *The Social Construction of What?* London: Harvard University Press.

Hacking, I. (1999b) *Mad Travellers: Reflections on the Reality of Transient Mental Illnesses.* London: Free Association Books.

Hacking, I. (1999c) 'When the trees talk back: a review of "Pandora's Hope" by Bruno Latour', *Times Literary Supplement,* 10 September, p. 13.

Hacking, I. (2002) *Historical Ontology.* Boston, MA: Harvard University Press.

Hacking, I. (2004) 'Between Michel Foucault and Erving Goffman: between discourse in the abstract and face-to-face interaction', *Economy and Society,* 33: 277–302.

Haggett, P. (1965) *Locational Analysis in Human Geography.* London: Arnold.

Hague, C. (2002) 'What is planning and what do planners do?', in P. Allmendinger, P. Prior and J. Raemakers (eds), *Introduction to Planning Practice.* Chichester: Wiley. pp. 1–20.

Halkier, B. (2001) 'Consuming ambivalences: consumer handling of environmentally related risks in food', *Journal of Consumer Culture,* 1: 205–24.

Hall, P., Gracey, H., Drewett, R. and Thomas, R. (1973) *The Containment of Urban England.* Vol. 1, *Urban and Metropolitan Growth Process or Megalopolis Denied.* London: Allen and Unwin.

Hamnett, C. (2003) 'Contemporary human geography: fiddling while Rome burns?', *Geoforum,* 34: 1–3.

Hannah, M. (1997) 'Imperfect panopticism: envisioning the construction of normal lives', in G. Benko and U. Strohmayer (eds), *Space and Social Theory: Interpreting Modernity and Postmodernity.* Oxford: Blackwell. pp. 344–59.

Haraway, D. (1991) *Simians, Cyborgs and Women: The Reinvention of Nature.* London: Routledge.

Haraway, D. (1997) *Modest Witness @ Second Millennium: FemaleMan Meets Oncomouse.* London: Routledge.

Harley, J. (1988) 'Maps, knowledge and power', in D. Cosgrove and S. Daniels (eds), *The Iconography of Landscape.* Cambridge: Cambridge University Press. pp. 277–312.

Harvey, D. (1989) *The Condition of Postmodernity.* Oxford: Blackwell.

Harvey, D. (1996) *Justice, Nature and the Geography of Difference.* Oxford: Blackwell.

Harvey, G. (1997) *The Killing of the Countryside.* London: Penguin.

Harvey, M., McMeekin, A. and Warde, A. (2004) *Qualities of Food.* Manchester: Manchester University Press.

Healey, P. (1993) 'The communicative work of development plans', *Environment and Planning B: Planning and Design,* 20: 83–104.

Healey, P. (1998) *Collaborative Planning.* London: Macmillan.

Hetherington, K. (1997) *The Badlands of Modernity: Heterotopia and Social Ordering.* London: Routledge.

Hinchliffe, S. (2001) 'Indeterminacy in-decisions: science, science policy and politics in the BSE (bovine spongiform encephalopathy) crisis', *Transactions of the Institute of British Geographers,* 26: 182–204.

Hindess, B. (1996) *Discourses of Power: From Hobbes to Foucault.* London: Sage Publications.

Hoch, C. (1994) *What Planners Do: Power, Politics, and Persuasion.* Chicago, IL: Planners Press.

Hodge, I. (2000) 'Countryside planning: from urban containment to sustainable development', in B. Cullingworth (ed.), *British Planning: 50 Years of Urban and Regional Policy*. London: The Athlone Press. pp. 92–115.

Irigaray, L. (1985) *The Sex That Is Not One*. Ithaca, NY: Cornell University Press.

Jackson, P. (1989) *Maps of Meaning*. London: Routledge.

Jacobs, J. (1961) *The Death and Life of Great American Cities*. New York, NY: Random House.

Jacobs, M. (1997) *Making Sense of Environmental Capacity*. London: Council for the Protection of Rural England.

Jakle, J. and Sculle, K. (1999) *Fast Food: Roadside Restaurants in the Automobile Age*. London: The Johns Hopkins University Press.

Jensen, O. and Richardson, T. (2004) *Making European Space: Mobility, Power and Territorial Identity*. London: Routledge.

Jim, C.Y. (2004) 'Green-space preservation and allocation of sustainable greening of compact cities', *Cities*, 21: 311–20.

Joyce, P. (2003) *The Rule of Freedom: Liberalism and the Modern City*. London: Verso.

Keith, M. and Pile, S. (1993) *Place and the Politics of Identity*. London: Routledge.

Knorr-Cetina, K. (1981) *The Manufacture of Knowledge: An Essay on the Constructivist and Contextual Nature of Science*. Oxford: Pergamon Press.

Kuhn, T. (1962) *The Structure of Scientific Revolutions*. Chicago, IL: University of Chicago Press.

Kurzweil, E. (1980) *The Age of Structuralism: Lévi-Strauss to Foucault*. New York, NY: Columbia University Press.

Lane, S. (2001) 'Constructive comments on D. Massey "Space-time, "science" and the relationship between physical geography and human geography"', *Transactions of the Institute of British Geographers*, 26: 243–56.

Lash, S. (1998) *Another Modernity/A Different Rationality*. London: Blackwell.

Latham, A. (2003) 'The possibilities of performance', *Environment and Planning A*, 35: 1901–6.

Latour, B. (1983) 'Give me a laboratory and I will raise the world', in K. Knorr-Cetina and M. Mulkay (eds), *Science Observed*. London: Sage Publications.

Latour, B. (1986) 'The powers of association', in J. Law (ed.), *Power, Action, Belief*. London: Routledge and Kegan Paul. pp. 264–80.

Latour, B. (1987) *Science in Action*. Milton Keynes: Open University Press.

Latour, B. (1988) *The Pasteurization of France*. Cambridge, MA: Harvard University Press.

Latour, B. (1991) 'Technology is society made durable', in J. Law (ed.), *A Sociology of Monsters: Essays on Power, Technology and Domination*. London: Routledge. pp. 103–30.

Latour, B. (1993) *We Have Never Been Modern*. Hemel Hempstead: Harvester Wheatsheaf.

Latour, B. (1994) 'On technical mediation – philosophy, sociology, genealogyy', *Common Knowledge*, 4: 29–64.

Latour, B. (1997a) 'On actor-network theory: a few clarifications', paper to the Actor Network and After Conference, Keele, July.

Latour, B. (1997b) 'Trains of thought: Piaget, formalism and the fifth dimension', *Common Knowledge*, 6 (3): 170–91.

Latour, B. (1999) *Pandora's Hope*. London: Harvard University Press.

Latour, B. (2004) *Politics of Nature*. London: Harvard University Press.

Latour, B. (2005) *Reassembling the Social*. Oxford: Oxford University Press.

Latour, B. and Crawford, T. (1993) 'An interview with Latour', *Configurations*, 2: 247–69.

Latour, B. and Woolgar, S. (1979) *Laboratory Life: The Social Construction of Scientific Facts.* Beverley Hills, CA: Sage Publications.

Laurie, I.C. (1979) *Nature in Cities: The Natural Environment in the Design and Development of Urban Green Space.* Chichester: Wiley.

Law, J. (1986) 'On power and its tactics: a view from the sociology of science', *Sociological Review*, 34: 1–38.

Law, J. (1994) *Organizing Modernity.* Oxford: Blackwell.

Law, J. (1997) 'Traduction/trahision – notes on ANT', Centre for Social Theory and Technology, University of Keele, Keele.

Law, J. (1999) 'After ANT: complexity, naming and topology', in J. Law and J. Hassard (eds), *Actor Network Theory and After.* Oxford: Blackwell. pp. 1–15.

Law, J. (2000) 'Objects, spaces, others', published by the Centre for Science Studies, Lancaster University at <http://www.comp.lancs.ac.uk/sociology/papers/law-objects-spaces-others.pdf>.

Law, J. (2002) 'On hidden heterogeneities: complexity, formalism and aircraft design', in J. Law and A. Mol (eds), *Complexities: Social Studies of Knowledge Practices.* Durham, NC: Duke University Press. pp. 116–41.

Law, J. (2004) 'Enacting naturecultures: a note from STS', published by the Centre for Science Studies, Lancaster University at <http://www.comp.lancs.ac.uk/sociology/papers/law-enacting-naturecultures.pdf>.

Law, J. and Hetherington, K. (1998) 'Materialities, spatialities, globalities', published by the Centre for Science Studies, Lancaster University at <http://www.comp.lancs.ac.uk/sociology/papers/law-hetherington-materialities-spatialities-globalities.pdf>.

Law, J. and Mol, A. (2000) 'Situating technoscience: an inquiry into spatialities', published by the Centre for Science Studies, Lancaster University, Lancaster LA1 4YN, UK, at <http://www.comp.lancs.ac.uk/sociology/papers/Law-Mol-Situating-Technoscience.pdf>.

Law, J. and Mol, A. (eds) (2002) *Complexities: Social Studies of Knowledge Practices.* Durham, NC: Duke University Press.

Law, J. and Urry, J. (2004) 'Enacting the social', *Economy and Society*, 33: 390–410.

Lee, N. and Brown, S. (1994) 'Otherness and the actor network: the undiscovered continent', *American Behavioural Scientist*, 37: 772–90.

Lee, N. and Stenner, P. (1999) 'Who pays? Can we pay them back?', in J. Hassard and J. Law (eds), *Actor-Network and After.* London: Routledge.

Lefebvre, H. (1991) *The Production of Space.* Oxford: Blackwell.

Lemke, T. (2001) '"The birth of bio-politics": Michel Foucault's lecture at the College de France on neo-liberal governmentality', *Economy and Society*, 30: 190–207.

Lévi-Strauss, C. (1964) *Mythologiques.* Paris: Plon.

Lévi-Strauss, C. (1969 [1949]) *The Elementary Structures of Kinship.* Boston, MA: Beacon Press.

Love, J. (1986) *McDonald's: Behind the Arches.* New York, NY: Bantam.

Lowe, P. (1977) 'Amenity and equity: a review of local environmental pressure groups in Britain', *Environment and Planning A*, 9: 35–58.

Lowe, P. and Goyder, J. (1983) *Environmental Groups in Politics.* London: Allen and Unwin.

Lowe, P., Clark, J., Seymour, S. and Ward, N. (1997) *Moralising the Environment.* London: UCL Press.

Luke, T. (1999) 'Environmentality as green governmentality', in E. Darier (ed.), *Discourses of the Environment.* London: Blackwell. pp. 121–51.

Lynch, K. (1960) *The Image of the City*. Cambridge, MA: Harvard University Press.
Lyotard, J.F. (1988) *The Differend*. Minneapolis, MN: Minnesota University Press.
Macnaghten, P. and Urry, J. (1998) *Contesting Natures*. London: Sage Publications.
Martin, R. (2001) 'Geography and public policy: the case of the missing agenda', *Progress in Human Geography*, 25: 189–210.
Massey, D. (1991) 'A global sense of place', *Marxism Today*, June: 24–9.
Massey, D. (1992) 'Politics and space/time', *New Left Review*, 196: 65–84.
Massey, D. (1998) 'Power-geometries and the politics of space-time', Hettner-Lecture, Department of Geography, University of Heidelberg, Heidelberg.
Massey, D. (1999a) 'Space-time, "science" and the relationship between physical geography and human geography', *Transactions of the Institute of British Geographers*, 24: 261–76.
Massey, D. (1999b) 'Spaces of politics', in D. Massey, J. Allen and P. Sarre (eds), *Human Geography Today*. Cambridge: Polity Press. pp. 279–94.
Massey, D. (2000) 'Entanglements of power: reflections', in J. Sharp, P. Routledge, C. Philo and R. Paddison (eds), *Entanglements of Power: Geographies of Domination/Resistance*. London: Routledge. pp. 279–87.
Massey, D. (2004) 'Geographies of responsibility', *Geografiska Annaler*, 86: 5–18.
Massey, D. (2005) *For Space*. London: Sage Publications.
Massumi, B. (2002) *Parables for the Virtual: Movement, Affect, Sensation*. Durham, NC: Duke University Press.
Matless, D. (1998) *Landscape and Englishness*. London: Reaktion.
Mazza, L. (1995) 'Technical knowledge, practical reason and the planner's responsibility', *Town Planning Review*, 66: 389–409.
McCormack, D. (2003) 'An event of geographical ethics in spaces of affect', *Transactions of the Institute of British Geographers*, 28 (4): 488–507.
McDonald's (1996) *McDonald's Corporation 1996 Annual Report*. Oak Brook, IL: McDonald's.
McLoughlin, J.B. (1969) *Urban and Regional Planning: A Systems Approach*. London: Faber.
McNay, L. (1994) *Foucault: A Critical Introduction*. London: Polity.
Meek, J. (2001) 'We do ron ron. We do ron ron: a review of "Fast Food Nation" by Eric Schlosser', *London Review of Books*, 23 (6): 3–6.
Merchant, C. (2003) *Reinventing Eden: The Fate of Nature in Western Culture*. London: Routledge.
Michael, M. (2000) *Reconnecting Culture, Technology and Nature: From Society to Heterogeneity*. London: Routledge.
Miles, S. (2003) *Michel Foucault*. London: Routledge.
Miller, P. and Rose, N. (1990) 'Governing economic life', *Economy and Society*, 19: 1–31.
Mol, A. and Law, J. (1994) 'Regions, networks and fluids: anaemia and social topology', *Social Studies of Science*, 24: 641–71.
Mol, A. and Law, J. (2002) 'Complexities: an introduction', in J. Law and A. Mol (eds), *Complexities: Social Studies of Knowledge Practice*. Durham, NC: Duke University Press. pp. 1–22.
Mullins, J. and James, P. (2004) 'Obesity claims are a US growth industry', *The Times*, 29 June, p. 6.
Mumford, L. (1961) *The City in History: Its Origins, Its Transformations, and Its Prospects*. London: Secker and Warburg.

Munro, R. (2004) 'Punctualizing identity: time and the demanding relation', *Sociology*, 38: 293–311.

Murdoch, J. (1997) 'Towards a geography of heterogeneous associations', *Progress in Human Geography*, 21: 321–37.

Murdoch, J. (1998) 'The spaces of actor-network theory', *Geoforum*, 29: 357–74.

Murdoch, J. (2001) 'Ecologising Sociology: actor-network theory, co-construction and the problem of human exemptionalism', *Sociology*, 35 (1): 111–33.

Murdoch, J. (2004) 'Putting discourse in its place: planning, sustainability, and the urban capacity study', *Area*, 36: 50–8.

Murdoch, J. and Abram, S. (2002) *Rationalities of Planning: Development versus Environment in Planning for Housing*. Aldershot: Ashgate.

Murdoch, J. and Lowe, P. (2003) 'The preservationist paradox: modernism, environmentalism and the politics of spatial division', *Transactions of the Institute of British Geographers*, 28: 318–32.

Murdoch, J. and Marsden, T. (1994) *Reconstituting Rurality: Class, Community and Power in the Development Process*. London: UCL Press.

Murdoch, J. and Miele, M. (2004) 'A new aesthetic of food? Relational reflexivity in the "alternative" food movement', in M. Harvey, A. McMeekin and A. Warde (eds), *Qualities of Food: Alternative Theoretical and Empirical Approaches*. Manchester: Manchester University Press. pp.156–75.

Murdoch, J. and Ward, N. (1997) 'Governmentality and territoriality: the statistical manufacture of Britain's "national farm"', *Political Geography*, 16: 307–24.

Murdoch, J., Lowe, P., Ward, N. and Marsden, T. (2003) *The Differentiated Countryside*. London: Routledge.

Neumann, R. (1998) *Imposing Wilderness: Struggles over Livelihood and Nature Preservation in Africa*. Berkeley, CA: University of California Press.

Norris, C. (1994) '"What is enlightenment?": Kant and Foucault', in G. Gutting (ed.), *The Cambridge Companion to Foucault*. Cambridge: Cambridge University Press. pp. 159–96.

Norton-Taylor, R. (1982) *Whose Land is it Anyway? Agriculture, Planning and Land Use in the British Countryside*. Wellingborough: Turnstone Press.

Nygard, B. and Storstad, O. (1998) 'De-globalisation of food markets? Consumer perceptions of safe food: the case of Norway', *Sociologia Ruralis*, 38: 35–53.

Osborne, T. and Rose, N. (1999) 'Governing cities: notes on the spatialisation of virtue', *Environment and Planning D: Society and Space*, 17: 737–60.

Osborne, T. and Rose, N. (2004) 'Spatial phenomenotechnics: making space with Charles Booth and Patrick Geddes', *Environment and Planning D: Society and Space*, 22: 209–28.

Owens, S. (1994) 'Land, limits and sustainability: a conceptual framework and some dilemmas for the planning system', *Transaction of the Institute of British Geographers*, 19: 439–56.

Owens, S. (1997) 'Giants in the path: planning, sustainability and environmental values', *Town Planning Review*, 68: 293–304.

Owens, S. and Cowell, R. (2002) *Land and Limits: Interpreting Sustainability in the Planning Process*, London: Routledge.

Parasecoli, F. (2003) 'Postrevolutionary chowhounds: food, globalization and the Italian Left', *Gastronomica*, 3: 29–39.

Patton, P. (1998) 'Foucault's subject of power', in J. Moss (ed.), *The Later Foucault*. London: Sage Publications. pp. 64–77.

Patton, P. (2000) *Deleuze and the Political*. London: Routledge.

Peet, R. (1998) *Modern Geographical Thought*. Oxford: Blackwells.

Peet, R. and Thrift, N. (1989) 'Political economy and human geography', in R. Peet and N. Thrift (eds), *New Models in Geography*. Vol. 2. London: Unwin Hyman. pp. 3–29.

Petrini, C. (1986) 'The Slow Food manifesto', *Slow*, 1: 23–4.

Petrini, C. (2003) *Slow Food: The Case for Taste*. New York, NY: Columbia University Press.

Philo, C. (1998) 'Animals, geography and the city: notes on inclusions and exclusions', in J. Wolch and J. Emel (eds), *Animal Geographies: Place, Politics and Identity in the Nature-Culture Borderlands*. London: Verso. pp. 51–71.

Philo, C. (2000) 'Foucault's geography', in M. Crang and N. Thrift (eds), *Thinking Space*. London: Routledge. pp. 205–38.

Pickering, A. (1994) *Constructing Quarks*. Chicago, IL: University of Chicago Press.

Pile, S. and Keith, M. (1997) *Geographies of Resistance*. London: Routledge.

Porter, T. (1995) *Trust in Numbers: The Pursuit of Objectivity in Science and Public Life*. Princeton, NJ: Princeton University Press.

Pottage, A. (1998) 'Power as an art of contingency: Luhmann, Deleuze, Foucault', *Economy and Society*, 27: 1–27.

Probyn, E. (2000) *Carnal Appetites*. London: Routledge.

Rajchman, J. (2001) *The Deleuze Connections*. London: MIT Press.

Ratcliffe, J. (1973) *An Introduction to Town and Country Planning*. London: Routledge and Kegan Paul.

Rawcliffe, P. (1998) *Environmental Pressure Groups in Transition*. Manchester: Manchester University Press.

Reader, J. (2004) *Cities*. London: Heinemann.

Resca, M. and Gianola, R. (1998) *McDonald's: una storia Italiana*. Varese: Baldini & Castoldi.

Richardson, T. (1996) 'Foucauldian discourse: power and truth in urban and regional policy making', *European Planning Studies*, 4: 279–92.

Rifkin, J. (1992) *Beyond Beef: The Rise and Fall of the Cattle Culture*. New York, NY: Dutton.

Ritzer, G. (1993) *The McDonaldization of Society*. London: Sage Publications.

Ritzer, G. (2004) *The McDonaldization of Society*. Revised New Century Edition. London: Pine Forge Press.

Robson, K. (1992) 'Accounting numbers as "inscription": action at a distance and the development of accounting', *Accounting, Organisations and Society*, 17: 685–708.

Rodwin, L. (2000) 'Images and paths of change in economics, political science, philosophy, literature, and city planning, 1950–2000', in L. Rodwin and B. Sanyal (eds), *The Profession of City Planning: Changes, Images and Challenges 1950–2000*. New Brunswick, NJ: Rutgers University. Centre for Urban Policy Research. pp. 3–26.

Rome, A. (2001) *The Bulldozer in the Countryside: Suburban Sprawl and the Rise of American Environmentalism*. Cambridge: Cambridge University Press.

Rose, G. (1993) *Feminism and Geography*. Cambridge: Polity Press.

Rose, N. (1991) 'Governing by numbers: figuring out democracy', *Accounting, Organisation and Society*, 16: 673–92.

Rose, N. (1999) *Powers of Freedom*. Cambridge: Cambridge University Press.

Rose, M. (2002) 'The seductions of resistance: power, politics and a performative style of systems', *Environment and Planning D: Society and Space*, 20: 383–400.

Rose, N. and Miller, P. (1992) 'Political power beyond the state: problematics of government', *British Journal of Sociology*, 42: 173–205.

Rouse, J. (1987) *Knowledge and Power: Towards a Political Philosophy of Science*. Ithaca, NY: Cornell University Press.

Rydin, Y. (1998) 'Land use planning and environmental capacity: reassessing the use of regulatory policy tools to achieve sustainable development', *Journal of Environmental Planning and Management*, 41, 749–65.

Rydin, Y. (2003) *Conflict, Consensus and Rationality in Environmental Planning: An Institutional Discourse Approach*. Oxford: Oxford University Press.

Sandercock, L. (2003) *Cosmopolis II: Mongrel Cities in the 21st Century*. Chichester: Wiley.

Sayer, A. (1984) *Method and Social Science*. London: Hutchinson.

Sayer, A. (2004) 'Seeking the geographies of power: review of "Lost Geographies of Power" by John Allen', *Economy and Society*, 33: 255–70.

Schlosser, E. (2001) *Fast-food Nation: The Dark Side of the All-American Meal*. New York, NY: Houghton Miflin.

Scott, M. (1969) *American Planning since 1890*. Berkeley, CA: University of California Press.

Scruton, R. (2004) *News from Somewhere*. Cambridge: Polity Press.

Serres, M. (1995) *The Natural Contract*. Ann Arbor, MI: The University of Michigan Press.

Serres, M. and Latour, B. (1995) *Conversations on Science, Culture and Time*. Ann Arbor, MI: The University of Michigan Press.

Shapin, S. (1995) 'Here and everywhere: sociology of scientific knowledge', *Annual Review of Sociology*, 21: 289–321.

Sharp, T. (1932) *Town and Countryside*. London: Allen and Unwin.

Sharp, J., Routledge, P., Philo, C. and Paddison, R. (eds) (2000) *Entanglements of Power: Geographies of Domination/Resistance*. London: Routledge.

Sheail, J. (2002) *An Environmental History of the Twentieth Century*. London: Palgrave.

Short, J.R., Fleming, S. and Witt, S. (1986) *House Building, Planning and Community Action*. London: Routledge and Kegan Paul.

Slater, D. and Ritzer, G. (2001) 'Interview with Ulrich Beck', *Journal of Consumer Culture*, 1 (2): 261–77.

Slow Food (1999) 'The official Slow Food manifesto', *Slow*, 13: 24.

Slow Food (2000) *The Ark of Taste and Praesidia*. Bra: Slow Food Editore.

Smart, B. (1994) 'Digesting the modern diet: gastro-porn, fast food and panic eating', in K. Tester (ed.), *The Flaneur*. London: Routledge. pp. 45–62.

Smart, B. (1999) 'Resisting McDonaldization: theory, process and critique', in B. Smart (ed.), *Resisting McDonaldization*. London: Sage Publications. pp. 1–27.

Smith, P. (2001) *Cultural Theory*. Oxford: Blackwell.

Smith, D. (2003) 'Deleuze and the liberal tradition: normativity, freedom and judgement', *Economy and Society*, 32: 299–324.

Soja, E. (1996) *Thirdspace: Journeys to Los Angeles and Other Real-and-imagined Places*. London: Blackwell.

Soper, K. (1995) *What Is Nature?* Oxford: Blackwell.

Soper, K. (2000) 'Realism, humanism and the politics of nature', *Radical Philosophy*, 102: 17–26.

Spivak, G.C. (1992) 'The politics of translation', in M. Barrett and A. Phillips (eds), *Destabilising Theory: Contemporary Feminist Debates*. Cambridge: Polity Press. pp. 177–200.

Star, S.L. (1991) 'Power, technology and the phenomenology of conventions: on being allergic to onions', in J. Law (ed.), *A Sociology of Monsters: Essays on Power, Technology and Domination*. London: Routledge. pp. 26–56.

Star, S.L. (1995) 'The politics of formal representation: wizards, gurus and organisational complexity', in S.L. Star (ed.), *Ecologies of Knowledge: Work and Politics in Science and Technology*. New York, NY: State University of New York Press. pp. 89–118.

Suchman, L. and Trigg, R. (1993) 'Artificial intelligence as craftwork', in S. Chaiklin and J. Lave (eds), *Understanding Practice: Perspectives on Activity and Context*. New York, NY: Cambridge University Press. pp. 144–72.

Sutton, P. (2000) *Explaining Environmentalism: In Search of a New Social Movement*. Aldershot: Ashgate.

Sutton, P. (2004) *Nature, Environment and Society*. London: Palgrave.

Tansey, G. and D'Silva, J. (1999) *The Meat Business*. London: Earthscan.

Taylor, N. (1998) *Urban Planning Theory Since 1945*. London: Sage Publications.

Tewdwr-Jones, M. and Allmendinger, P. (1998) 'Deconstructing communicative rationality: a critique of Habermasian collaborative planning', *Environment and Planning A*, 30: 793–810.

Thomas, K. (1984) *Man and the Natural World*. London: Penguin.

Thrift, N. (1996) *Spatial Formations*. London: Sage Publications.

Thrift, N. (1999) 'Steps to an ecology of place', in D. Massey, J. Allen and P. Sarre (eds), *Human Geography Today*. Cambridge: Polity Press. pp. 295–322.

Thrift, N. (2004a) 'Summoning life', in P. Cloke, P. Crang and M. Goodwin (eds), *Envisioning Human Geographies*. London: Arnold. pp. 81–103.

Thrift, N. (2004b) 'Intensities of feeling: towards a spatial politics of affect', *Geografiska Annaler*, 86: 57–78.

Thrift, N. and Dewsbury, J.D. (2000) 'Dead geographies – and how to make them live', *Environment and Planning D: Society and Space*, 18: 411–32.

Torquati, B. and Frascarelli, A. (2000) 'Relationship between territory, enterprises, employment and professional skill in the typical products sector', in B. Sylvander, D. Barjolle and F. Arfini (eds), *The Socio-Economics of Origin-Labelled Products in Agrifood Supply Chains*. Le Mans: INRA.

Tudor, A. (1999) *Decoding Culture*. London: Sage Publications.

Unwin, R. (1910) *Town Planning in Practice: An Introduction to the Art of Designing Cities and Suburbs*. London: Fisher Unwin.

Urban Task Force (1999) *Towards Urban Renaissance*. London: Department of Environment, Transport and the Regions.

Vidal, J. (1997) *McLibel: Burger Culture on Trial*. London: Macmillan.

Vigar, G., Healey, P., Hull, A. and Davoudi, S. (2000) *Planning, Governance and Spatial Strategy in Britain: An Institutionalist Approach*. London: Macmillan.

Wackernagel, M. and Rees, W. (1996) *Our Ecological Footprint: Reducing Human Impact on the Earth*. Gabriola Island, BC: New Society Publishers.

Ward, S. (1994) *Planning and Urban Change*. London: Paul Chapman Publishing.

Whatmore, S. (1997) 'Dissecting the autonomous self: hybrid cartographies for a relational ethics', *Environment and Planning D: Society and Space*, 15: 37–53.

Whatmore, S. (1999) 'Hybrid geographies: rethinking the "human" in human geography', in D. Massey, J. Allen and P. Sarre (eds), *Human Geography Today*. Cambridge: Polity Press. pp. 22–40.

Whatmore, S. (2002) *Hybrid Geographies: Natures, Cultures, Spaces*. London: Sage Publications.

Whatmore, S. and Hincliffe, S. (2003) 'Living cities: towards a politics of conviviality', presentation to the 'Technonatures: Environments, Technologies and Spaces in the 21st Century' symposium, Goldsmiths College, London, June.

Williams, R. (1973) *The Country and the City*. London: Chatto and Windus.

Winter, M. (1996) *Rural Politics*. London: Routledge.

Wolch, J. (2002) 'Anima urbis', *Progress in Human Geography*, 26: 721–42.

World Commission on Environment and Development (WCED) (1987) *Our Common Future*. London: Routledge.

Wu, F. and Webster C.J. (1998) 'Simulation of land development through the integration of cellular automata and multicriteria evaluation', *Environment and Planning B: Planning and Design*, 25: 103–26.

Wynne, B. (1996) 'SSK's identity parade: signing-up, off-and-on', *Social Studies of Science*, 26: 357–91.

Zammito, J. (2004) *A Nice Derangement of Epistemes: Post-positivism in the Study of Science from Quine to Latour*. Chicago, IL: University of Chicago Press.

Index